EARTH SCIENCES IN THE AGE OF THE SATELLITE

JEAN POUQUET

University of Aix-Marseille

EARTH SCIENCES IN THE AGE OF THE SATELLITE

D. REIDEL PUBLISHING COMPANY

DORDRECHT-HOLLAND / BOSTON-U.S.A.

LES SCIENCES DE LA TERRE A L'HEURE DES SATELLITES

First published in 1971 by Presses Universitaires de France, Paris

Translated from the French by Isabel A. Leonard

Linguistic Systems Inc., Cambridge, Massachusetts

Library of Congress Catalog Card Number 73–94454

ISBN 90 277 0437 6

Published by D. Reidel Publishing Company,
P.O. Box 17, Dordrecht, Holland

Sold and distributed in the U.S.A., Canada and Mexico
by D. Reidel Publishing Company, Inc.
306 Dartmouth Street, Boston,
Mass. 02116, U.S.A.

Printed in The Netherlands by D. Reidel, Dordrecht

TABLE OF CONTENTS

APPENDIXES

LIST OF PLATES

BY WAY OF A PREFACE

Renaissance man "... discovered many a mechanical marvel.... The achievement of the astronauts... opened up comparable prospects to the men of today, but of infinitely wider scope".

C. LUCET,
French Ambassador to the United States.*

"Any future... must inevitably pass through the channel of combined disciplines... (from which) will arise a humanization of state-of-the-art technology, and updating of methods of Earth Science."

Author unknown.**

"It is difficult to say what is impossible, for the dream of yesterday is the hope of today and reality tomorrow."

ROBERT GODDARD,
American physicist. This phrase has become the symbol of NASA.

* Is there a crisis of the spirit?, *Bul. Soc. Prof. Fr. en Amér.* (1969) p. 9.
** *La recherche spatiale* [Space Research] (May 1969) p. 15.

FROM GALILEO TO ALDRIN AND ARMSTRONG

In 1610, Galileo observed the surface of the Moon through the lens which bears his name and announced that, contrary to official opinion, its surface was irregular and not smooth. We now know that this observation – and many others of Galileo – was a correct one, but the opposition that always arises against research too far ahead of its time resulted in his condemnation.

From 1610 to 1960, the Earth was the observer. In 1960, it was observed: in April of 1960, the satellite Tiros 1 transmitted the first picture of our planet as it is seen from space.* On 20 July, 1969, the American astronauts Aldrin and Armstrong stepped out of the Eagle onto the surface of the Moon, and 'confirmed' the correctness of the statements made three centuries before. Since then, not a day has passed without bringing us pictures of our globe from space (see Plate A, Photo 1, opposite page 59) and though more infrequently, pictures of other planets.

Conquest of space has always been one of the dreams of mankind. Poets and writers have let their imagination run wild. Even a politician, speaking to the House of Representatives, said: "The American nation is growing rapidly.... This is our pride and our danger, our weakness and our strength... let us unite and... conquer space".**

Some forty years later, 1858†, the Frenchman G.-F. Tournachon began the conquest of space in his own manner: he took photographs of Paris from a balloon. In 1862, General G. B. McClellan used the same procedure during the American Civil War. The services rendered by use of aerial photography during the two world wars are well known to us, as well as that of infrared photography during the second world war.

All of these photographic techniques are forms of remote detection. Detecting something from a distance is as old as life itself. Our nose, ears, the sense of touch and, especially that of sight are among the best detectors. As soon as an object is perceived, man feels the need to hold his impression, either to make it last or to analyze it. One more stage had to be passed: that of inventing more perfect instruments and, concomitantly, holding the impressions perceived by these instruments in order to make them useful to man.

* We should define the altitude at which space begins. Actually, in 1946, V2 rockets were taking photographs from an altitude of about 50 km.

** John C. Calhoun, representative of South Carolina: at that time (1817) Secretary of State for War.

† It is well to point out that in 1840, Draper (United States), then Rutherford, in 1850, took the first photographs (daguerreotypes) of the Moon. In 1845, in France, Fizeau and Foucault photographed the Sun.

There is nothing new about this: for a very long time the police, museums, and banks have possessed equipment for seeing 'the invisible.' X-rays, infrared, and ultraviolet radiation are used every day to analyze the brushwork of famous artists and detect forgeries and, on a more mundane level, identify forged checks.

At the end of the Second World War, and more recently in Vietnam, certain military units have had weapons permitting them to 'see' at night with an invisible eye. These weapons use an infrared sensor that picks up the heat radiated by the enemy's body, which is warmer than its environment.

All of these techniques are part of the arsenal of remote detection, a term which appeared in the scientific vocabulary around 1960. Many authors seem to use the term for phenomena perceived from a platform in lunar or terrestrial orbit. I neither intend to discuss the validity of this restriction nor refute or approve it as such a point seems academic. However, until about 1960, the devices used for remote detection were fairly limited and had been slow in developing. Here I speak mainly of photography, whether black and white, color, aerial, or oblique. Improvements were made in the quality of emulsions and in analysis procedures such as plotting and densitometry.

It is my belief that some psychological shock resulted from the first images transmitted from a platform orbiting above the stratosphere. These pictures were very rough, due to the rudimentary TV cameras on board the satellite and ground receiving equipment, but it had been proved that one could transmit pictures in the visible range – *T*elevision, and in *I*nfra *R*ed, from an *O*bservatory on board a *S*atellite. These initials spell out the name of *TIROS* (*T*elevision *I*nfra *R*ed *O*bservation *S*atellite).

I wrote *psychological shock* quite intentionally, for that is what is was. Inspired by the perspectives offered by the future, researchers set to work at a feverish pace. Improvement followed improvement in television, infrared detectors, and other means of transmitting, receiving, decoding, and information processing.

If we pause for a moment to reflect, we can see that nothing is new here. Highly developed television techniques had been around for a long time, infrared detectors even longer, but it was not until the blast-off of Sputnik 1 and later satellites that their dormant possibilities could be truly exploited.

All instruments placed on board a satellite are first subjected to lengthy testing, first in the laboratory, then in aircraft. Conversely, it is perfectly possible for airplanes to carry out missions now performed by spacecraft. This is becoming quite common in the United States, at least. One curious point arises here: while the expressions 'remote sensing' or 'remote detection' are avoided for traditional aerial photography missions, they are used when the airplane carries radar or infrared detectors, and when multispectral photography, even if it is only in the visible range, is either the principal or a minor function of the mission concerned.

I hope these lines will assist the reader in a better understanding of this book. Although the title incorporates the word 'satellite', I am not going to confine myself exclusively to spacecraft. I am going to talk about remote detection in general, always remembering the role and achievements of satellites: this book is written with the scientific payload of spacecraft in mind. I am not going to stop at achievements up to 1974 as the near future is known and experiments in aircraft have already been made. If it were not for financial problems and the time needed to produce 1979, and 1980 satellites, they could be placed in orbit now, as the technical difficulties have been resolved.

Earth science specialists still have a few short years left to familiarize themselves with these new research techniques. Here, I advise the reader to reread the quotations at the beginning of the book. The second quotation, more 'down to earth,' can help us come up to the level demanded by today's science. I am distressed to see that a good number of our students in geography, geology, and soil science are still ignorant of the *basic knowledge of our era*, by which I mean the *mathematical and physical sciences*, essential to present-day research in any field.

1. Some Basic Observations

(1) Some people will doubtless be astonished not to find the pseudoscientific jargon so dear to many; consequently, I will confirm my total scientific ignorance by consistently using as simple a vocabulary as possible.

(2) Faithful to the principle of linguistic clarity, I have allowed myself some latitude in terminology. For example, instead of *Reststrahlen* I will use dip, or local minimum. In electromagnetic radiation, the German word means a sudden dip between two close bands of the spectrum.

(3) I have eliminated the mathematical approach in favor of 'directions for use'. However, I must counsel non-physics students to refresh and improve their knowledge of mathematics and physics so that, as they progress far beyond the contents of this book, they will be in a position to pursue their studies in this exciting field of space science.

(4) Throughout this book I will use one or other of the sets of abbreviations below:

(a) *Lengths:*
1 mm = 1000 μm, or μ (micrometer, micron);
1 μm (or μ) = 1000 nanometers, nm, or millimicrons, mμ;
1 μm = 10000 Ångströms. 1 mμ (or nm) = 10 Å.

(b) *Frequencies:*
Hertz, Hz, or cycles, c (1 Hz = 1 c/second);
Kilohertz, kHz = 1000 cycles, 10^3 Hz;
Megahertz, MHz = 1 million c, 10^6 Hz;
Gigahertz, GHz = 1 billion c, 10^9 Hz (or kilomegahertz, kMHz).

(c) *Wavelengths:* The usual terminology will be employed.

(d) *Angles:* You will read the angular value in degrees or, more often, the linear value in radians.

This is a good place to recall that 1 radian, the arc of a circle equal to the radius, is subtended by an angle of $180°/\pi = 57.2958°$. One degree subtends $\pi/180 = 0.01745$ rad.

(5) *Frequency-wavelength correspondences:* wavelength is the velocity of light (3×10^5 km/second) divided by the frequency.*

Wavelengths expressed in microns (μm):

300 000 000 000 000/Hz (3×10^{14}/Hz)
300 000 000 000/kHz (3×10^{11}/kMz)
300 000 000/MHz (3×10^{8}/MHz)
300 000/GHz (3×10^{5}/GHz)

Wavelengths expressed in millimeters:

300 000 000 000/Hz (3×10^{11}/Hz)
300 000 000/kHz (3×10^{8}/kMz)
300 000/MHz (3×10^{5}/MHz)
300/GHz (3×10^{2}GHz)

Wavelengths expressed in centimeters:

30 000 000 000/Hz (3×10^{10}/Hz)
30 000 000/kHz (3×10^{7}/kMz)
30 000/MHz (3×10^{4}/MHz)
30/GHz

Wavelengths expressed in meters:

300 000 000/Hz (3×10^{8}/Hz)
300 000/kHz (3×10^{5}/kHz)
300/MHz (3×10^{2}/MHz)
0.3/GHz

Ex. 1. What is the wavelength of the frequency 750 kHz (read on my radioset, AM)?
In meters $300\,000/750 = 400$ meters.

Ex. 2. A European station broadcasts on 500 m wavelength. How should I dial on my American set?
$300\,000/500 = 600$ kHz (usually only 6 on the dial);
(frequency = light speed/wavelength).

* See A. G. Pacholczyk, *Radio Astrophysics*, Freeman, New York 1970. Figure A3, p. 236: an excellent chart of frequency-wavelength conversions.

PART 1

BASIC PRINCIPLES OF REMOTE DETECTION

Part 1 sets out to recall to non-physics readers some technical essentials from their classroom days. These, as we shall see, are traditional, well known concepts: the electromagnetic spectrum as a whole, and its visible, infrared, and microwave components.

One aspect, computers, will not be considered, although they are the basis of space achievements and successful remote detection. All of the calculations and maps are performed and drawn with their aid; without computers, we would still be at the 'medieval' stage of the post-second-world-war days.

Before 'turning the page' I would like to pay homage to the mathematicians and physicists who have at ever-increasing speeds improved and are still improving Pascal's ancient calculating machine. Their defect is obvious: these computers can only add 0 and 1 (current off or on) but the operations broken down into these tiny elements are carried out at the speed of light, and we see no limit to the complexity of the problems that can be put to them.

I have used computers from their infancy and continue to do so, seeing new 'miracles' from week to week. Without these machines, this book would never have seen the light of day.

THE ELECTROMAGNETIC SPECTRUM

Remote detection consists of picking up signals reflected or emitted by water, rocks, plants, or clouds; whether perceived or not by the senses of man. All objects reflect to a greater or lesser extent the radiation emitted by the Sun, or any other source of energy. All objects, provided their temperature is above absolute zero (0 K, −273 °C) emit radiations part of which can be perceived by special instruments.

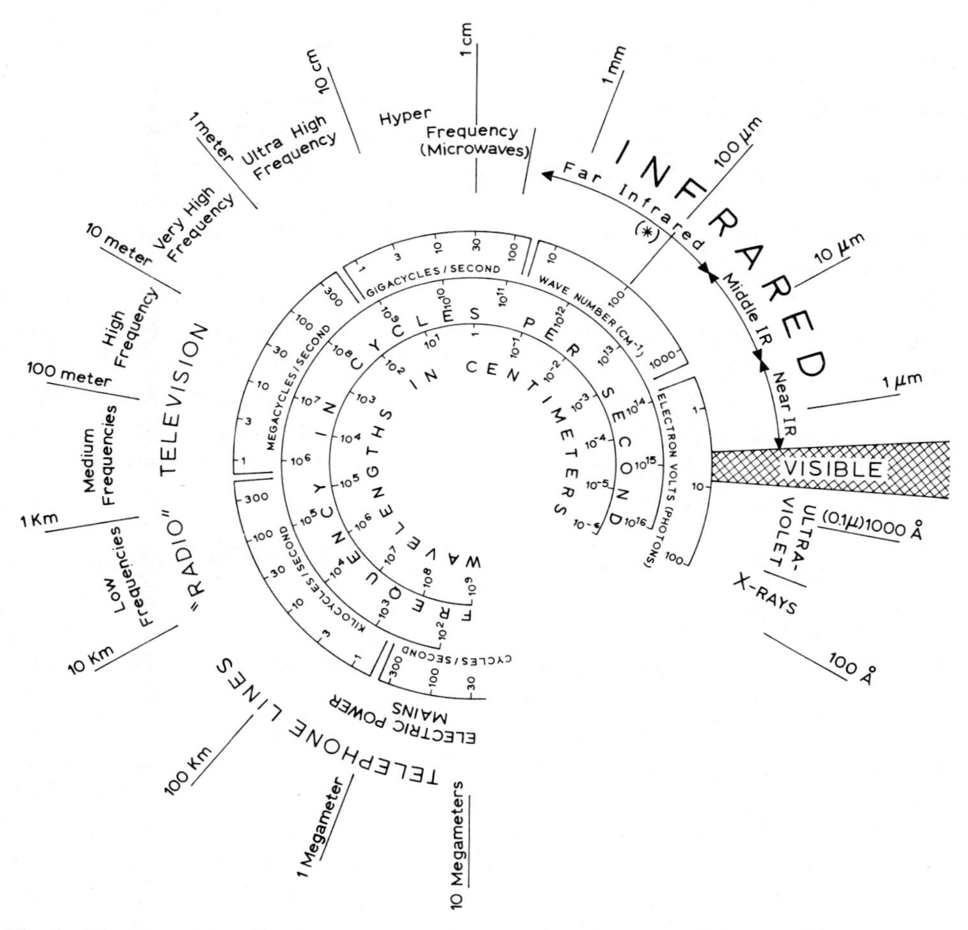

Fig. 1. The electromagnetic spectrum. – The boundary for far infrared, generally set at 1 mm, has been set here at a longer wavelength. (Hadni, 1969).

The Sun is an emitter par excellence of short wavelength electromagnetic radiations, and objects re-emit part of the energy absorbed on a longer wavelength. The electromagnetic spectrum is characterized by its continuity (*continuum*) and by its nature; i.e., the number of individual signals emitted in one second and, thus, the linear distance between two consecutive signals. Frequency (number of signals per second) and *wavelength* (distance between two signals, or more precisely, between two successive cycles) are the means of identifying electromagnetic waves travelling at the speed of light, or 3×10^5 km per second.

Figure 1 shows the reciprocal relationships between frequency and wavelength, the wavelength being the speed of light divided by frequency. If a single signal per second, or a single *cycle* or *Hertz*, characterized a wavelength, it would be 3×10^5 km long. We can thus attempt a simplified classification of the electromagnetic continuum, based on the practical usage of the various bands of the spectrum.

1. Spectral Bands not Used by Earth Sciences

These are the *frequencies less than 3 GHz*. Among these, we may set aside those for the electric power line networks (10 to 100 Hz) and the telephone networks (300 Hz to 30 kHz). The bands of the spectrum reserved for television and radio are universally divided into *low frequency* (10^5 Hz–300 kHz, or 3000 m–1000 m); *medium frequency* (10^6 Hz–3 MHz); 300 m–100 m); *high frequency* (10^7 Hz–30 MHz; 30 m–10 m); *very high frequency* (10^8 Hz–300 MHz; 3 m–10 m, and *ultra high frequency* (10^9 Hz–3 GHz; 30 cm–10 cm).

We may recall that our so-called FM (*frequency modulation*) sets, as well as the VHF television channels, operate in very high frequency, while AM (*amplitude modulation*), such as the old wireless, occupies the low and medium frequencies (long wave and 'little wave') and high frequency ('short wave') bands. We should also recall that ultra high frequency is being used more and more for television transmitters (UHF channels).

2. Bands of the Spectrum Usable and Partially Used in Earth Sciences

2.1. Microwaves (hyperfrequencies)

Microwaves should more properly be called centi-millimetric waves. With microwaves we approach the domain of remote detection; however, this does not mean that lower frequencies will not be exploited in the future despite growing galactic noise. The scientific literature does not agree on the limits of the spectrum bands, so we should not be surprised to see some overlaps. Centi-millimetric waves are the best known, since radar operates in this region of the spectrum.

Many scientists consider microwaves to extend from 10^9 to 10^{11} Hz (30 cm–3 mm). Others push these upper and lower limits to 300 MHz (1 m) and 300 GHz (1 mm); the latter limits are those most commonly observed in remote detection. As we shall see later, microwaves are extremely promising, since they are not 'blocked' by cloud, mist, fog, and even rain if the wavelength is long enough. Progress has already been made, and some satellites launched from 1972 on, carry passive receivers, but with low ground resolution.

2.2. Infrared

For some, infrared begins at 3 mm, for most at 1 mm. Everyone is agreed on 0.7–0.75 μm as the upper limit of the visible part of the spectrum.

(a) *Far infrared*, from 3 mm (10^{11} Hz)* or 1 mm (300 GHz) to 100 μm (3000 GHz). This band of the spectrum has never been used for remote detection and seems unlikely to be except for the lower frequencies associated with centi-millimetric wavelengths.

(b) *Middle infrared* between 100 μm (3000 GHz) and 3 μm (10^{14} Hz). This band is intensively exploited from 14 to 3 μm (thermal infrared).

(c) *Near infrared*, below 3 μm, up to the visible range. This band, somewhat neglected until 1969, is being more and more extensively used (reflected infrared).

2.3. Visible band of the solar spectrum

This is the part most familiar to us, extending from 700 to 400 nanometers (7000–4000 Å; 0.7–0.4 μm). The visible band is presently being favored by renewed interest, since in multispectral photography we can 'go down' as far as 300 mμ (ultraviolet) and 'go up' as far as 900 mμ (near infrared).

2.4. Ultraviolet and very short wavelengths

In this region atmospheric opacity interferes with reception of these spectrum bands, including ultraviolet, X-rays, and gamma rays, thus prohibiting their use.

The ultraviolet offers numerous opportunities for use but apparently, satellites planned up to 1980 will not carry instruments operating in these bands, with the exception of photographic emulsions sensitive to the very near ultraviolet which could be used by astronauts or personnel in the Skylab and Spacelab manned orbiting platforms, with recoverable shuttles plying to and from earth.

To sum up, we are interested only in these bands: 0.4–0.7 μm (visible); 0.7–3 μm (near infrared); 3–14 μm (middle infrared) and microwaves.

Table I sums up the information given so far.

* See A. Hadni, *The far infrared, 'SUP – Le Physicien'*, collection, No. 2, Paris, P.U.F., 1969, for the limit to 3 mm.

TABLE I

Utilization of spectral bands[a]

Frequency	Wavelength	Current designation	Utilization
10^1 Hz	30000 km		Electric lines, telephone
10^2 –	3000		
10^3 –	300		
10^4 –	30		
10^5 Hz	3000 m	Low frequency	
10^6 –	300	Medium frequency	Radio, television
10^7 –	30	High frequency	
10^8 –	3	Very high frequency	
10^9 –	0.3	Ultra high frequency	
3×10^8 Hz	100 cm		
10^9 –	30	Active (Radar) or	EARTH SCIENCES
10^{10} –	3	passive microwaves[b]	Growing usage
10^{11} –	0.3		
10^{12} Hz	300 μm		
10^{13} –	30	Infrared	Used below 15 μm
10^{14} –	3		
3×10^{14}–	1		
414000 GHz	0.7 μm	Visible	Massive use
750000 –	0.4		
10^{15} Hz	0.3 μm	Ultraviolet	Starting to be used
Over 10^{15} Hz	Less than 0.3 μm	X-Rays gamma rays	UNUSABLE (Police detection, Medical, etc.)

[a] In the world of telecommunications, the following classification is normally used:

Very low frequencies	10 kHz – 30 kHz	(30 km– 10 km)
Low frequencies	30 kHz –300 kHz	(10 km– 1 km)
Medium frequencies	300 kHz – 3 MHz	(1 km–100 m)
High frequencies	3 MHz– 30 MHz	(100 m – 10 m)
Very high frequencies (VHF)	30 MHz–300 MHz	(10 m – 1 m)
Ultra high frequencies (UHF)	300 MHz– 3 GHz	(1 m – 10 cm)
Hyper frequencies (or super high frequencies)	3 GHz– 30 GHz	(10 cm– 1 cm)

Actually, the hyper frequencies (microwaves) 'stretch' up to 1 m (300 MHz) and down to 1 mm (300 GHz).

Over 30 MHz (wavelengths smaller than 10 m), the radio signals are *not* reflected by the D, E, F^1, F^2, G layers of the ionosphere as the smaller frequencies (longer wavelengths) are a phenomenon intensively used for long distance telecommunications. As a consequence, frequencies greater than 30 MHz must be received in direct transmission between the broadcaster and the receiving stations. As the ionosphere extends between 80 km and 400 km in altitude, and the satellite is in orbit over the ionosphere (sometimes within it), it is impossible to use the 'normal' wavelengths. Consequently, for all communications between the Earth and the satellite, 'play back' of the recorded magnetic tapes, 'interrogations', 'orders' sent to the satellite..., the only wavelengths available belong to the VHF, UHF and hyper frequencies: the receiving station and the satellite must be in direct view.

[b] The P radar band operates from 220 to 390 MHz (1.33 m–77 cm).

CHAPTER II

THE VISIBLE AND NEAR INFRARED OF THE ELECTROMAGNETIC SPECTRUM

Since some objects absorb more energy than others, solar radiation is partially and unequally reflected by different objects. The eye and photographic film register only reflected radiation; the remainder of the solar energy is absorbed by objects, atmospheric gases, and/or refracted by the upper atmospheric boundary and/or diffracted by the atmosphere itself.

That part of the solar energy reaching the ground and that part reflected are more or less scattered; the scattering being at a maximum in the short wavelengths, from about 0.4 to 0.5 μm, in the violet and blue bands, and at a minimum in the orange and red regions of the visible spectrum. Roughly speaking, the scattering of visible radiation is inversely proportional to the fourth power of the wavelength. At 400 mμ (violet), scattering is about four times as great as at 600 mμ (orange). Thus, when taking photographs through a thick atmospheric layer (from an aircraft or satellite) one must avoid scattering zones that attenuate contrast and make objects somewhat blurred. This can be done easily by using appropriate filters.

At this point, it is useful to note that radiometers, spectrometers, interferometers, and television cameras by no means constitute the complete range of apparatus used for remote detection. We have all seen the photographic records brought back to earth by the Gemini and Apollo astronauts. In addition, instrumentation for photographic shots is built into TV cameras carried on board unmanned satellites. Finally, orbiting platforms such as Skylab and Spacelab permit intensive utilization of photographic emulsions. It is thus necessary to devote a few pages to this already ancient technique, now undergoing rapid rejuvenation.

1. The Basic Principles of Photography

1.1 Time and Season

Sensitive surfaces record only reflected solar energy between 350 mμ and 950 mμ, pushing into the ultraviolet and infrared. The energy registered on the film is directly dependent on the quantity and spectral quality available of solar energy, which leads us to the concept of the Sun's angle above the horizon. We all know that daylight changes as the day advances according to the slope of the Sun's rays. When the Sun is low on the horizon, the blue band (from 0.4 to 0.5 μm) is extremely scattered due to the greater thickness of the atmospheric layer: thus the Sun appears reddish. It is thus recommended that the early morning and late evening hours be avoided for photography. The photographs most useful to Earth scientists are taken around midday when illumination and spectral quality are at their best. We can now understand why the television images transmitted by the Nimbus satellites are of such exceptional quality: these satellites are scanning the various points of the globe at local times between 11:00 and 13:00, between about 50° latitude north or south, and earlier or later at higher latitudes.

The ideal season of the year varies according to the subject being photographed. In April, in the northern hemisphere, it is fairly easy to distinguish fertile soils since plant life takes hold rapidly while, if we wish to discriminate between types of vegetation, autumn is preferable. After a rainfall, when the air has been washed clear of suspended dust, higher contrast plates can be taken. A slight breeze helps to show up water pollution, probably because of differences in density.

These considerations must be borne in mind when we are working on satellite records, especially television plates, and data from the near infrared, up to around 2.5 μm.

1.2. Use of color filters

In general terms, a colored filter attenuates the influence of the color complementary to it, and favors the spectral band corresponding to its own color.

Table II shows some of the characteristics of filters currently used (the numbers are for Wratten filters) while Figure 2 presents, in an unusual but pedagogically practical form, the relationships between spectrum bands and filters; the bands are identified by their wavelengths and the corresponding colors.

In satellites, the filters are placed on the TV camera lenses. Filter No. 12 is the most frequently used. The TV photographs from Nimbus 3, due to the yellow filter known as 'minus blue filter', are characterized by the striking contrast of the cloud forma-

TABLE II

Characteristics of some colored filters

Filter type	Wavelength attenuated + or −	Favored wavelength	Exposure indexes	
			Panchromatic film	Infrared film
No. 12 (yellow)	Less than 0.3 μm and 0.34 to 0.49 μm	0.31 to 0.34 μm and over 0.5 μm	× 2	× 1.5
No. 15G (orange)	Less than 0.31 μm and 0.31 to 0.51 μm	0.31 to 0.32 μm and over 0.53 μm	× 2	× 1.5
No. 25A (red)	Less than 0.58 μm	Over 0.62 μm	× 4	× 2
No. 57A (green)	Less than 0.4 μm and 0.64 to 0.68 μm	0.42 to 0.62 μm and 0.64 to 0.7 μm	× 3	Not recommended
No. 58 (green)	Less than 0.47 μm and over 0.62 μm etc.	0.48 to 0.62 μm 'peak' at 5.3 μm etc.	× 2.5	Not recommended

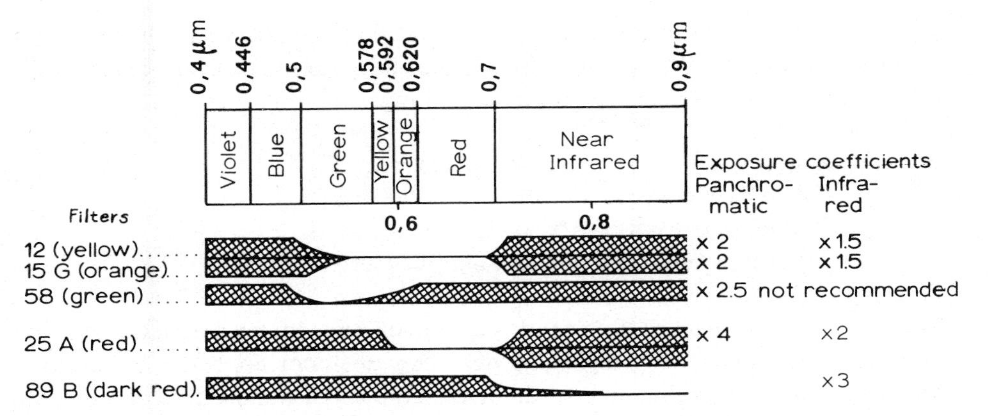

Fig. 2. Some Wratten filters used in multispectral photography. – Crosshatched zones: bands of the spectrum eliminated by the filters.

tions which facilitates the work of meteorologists. On the *ATS 3* Satellite (*A*pplications *T*echnology *S*atellite), three cameras are provided with different filters; on the ground, the three plates are then color restored using the combination of three spectral bands. We are thus approaching one of the most interesting subjects in the future of remote detection: multispectral photography.

2. Multispectral Photography

(See Plate B, opposite page 61)

This may not be the place to deal with the various photographic emulsions. Moreover, infrared, black-and-white or color films do not deserve special explanations. We should remember, however, that the power of infrared-sensitive emulsions does not, practically speaking, extend beyond 0.9 μm. It is not advisable to expose infrared film without a filter, as in this case, results are far superior with normal black-and-white or color panchromatic films.

In multispectral photography we should not attempt to reproduce the subtly shaded colors and tints of nature. The important thing is *to bring out this or that phenomenon, using a given band of the spectrum*. For example, a green filter enables us to better penetrate water, and discern the sea depths just offshore. Apparently healthy but actually withering trees show up in the 0.32–0.38 μm band as having lighter shades; in the 0.7–0.9 μm band, instead of being red as would healthy plants, they show up as pale red or orange, or shades even further removed from red.

Let us take the case of infrared color film. On the plates, the healthy vegetation shows up red, but objects with the same shade of green as neighboring plants appear to be blue, whence the name *photo-camouflage* given to this simple technique during the Second World War.

Since satellites are being pushed somewhat into the background of this chapter, we will confine this discussion to the *spectral bands used by the Gemini and Apollo astronauts*; these bands were retained by the *ERTS* satellites (*E*arth *R*esources *T*echnology *S*atellites). By way of simplification, but also for practical reasons, the spectrum bands up to 900 m are termed as follows:

Blue (violet + blue), from 400 to 500 mμ;
Green (green + yellow), from 500 to 600 mμ;
Red (orange + red), from 600 to 700 mμ;
Infrared, from 700 to 900 mμ (and beyond).

We already know that the blue band is always eliminated when aerial or satellite photographs are taken from high altitudes. This does not apply on the ground, where scattering is negligible due to the lesser amount of atmosphere between the camera and the object photographed.

On most of the Apollo satellites, not including Apollo 9, the following spectrum

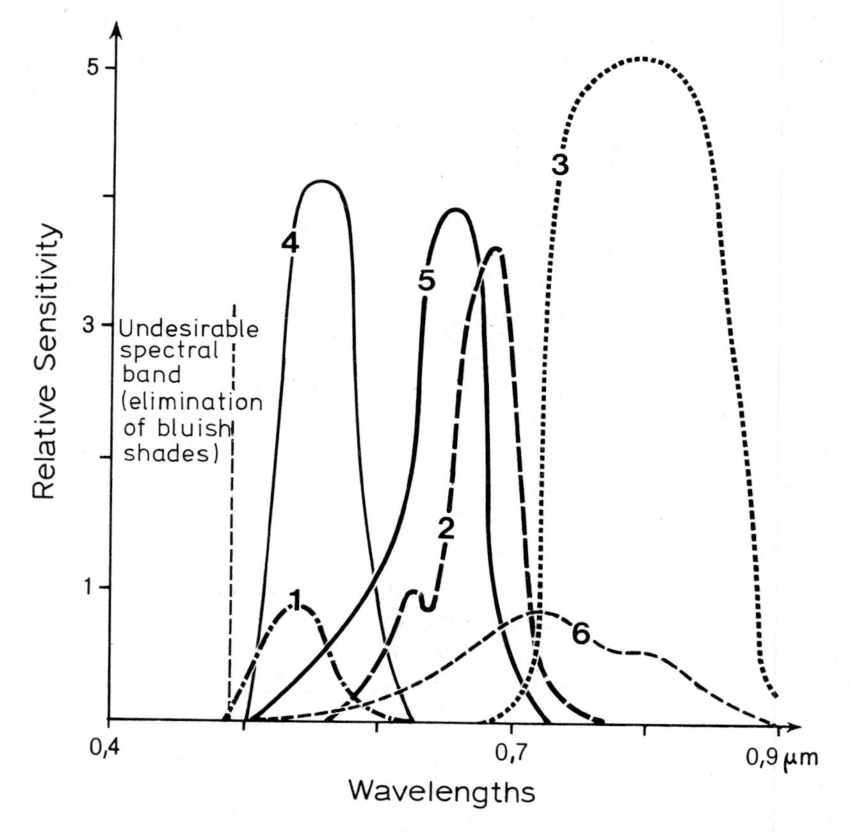

Fig. 3. Comparison of relative sensitivity of photographic emulsion associated with Wratten filters (15, 25A, 58, and 89B) – 1. Panatomic X and filter 58 (green); 2: Panatomic X and filter 25A (red); 3: Infrared black and white and filter 89B (dark red); 4, 5, and 6: Ektachrome infrared, filter 15 (orange), 4: yellow dominant; 5: magenta dominant; 6: cyan dominant (sensitive layers before exposure).

bands were used: 0.47–0.61 μm (green); 0.59–0.72 μm (red) and 0.72–0.9 μm (infrared). For the ERTS (Earth Resources) satellites, the above scheme was altered slightly; *green*, filter No. 58, from 475 to 575 mμ; *red*, filter No. 25A, from 580 to 680 mμ, and infrared, filter No. 89B, from 690 to 830 mμ.

Figure 3, grossly exaggerated, shows the sensitivity of photograph emulsions relative to each of these filters. We can immediately see that the sensitivity peaks correspond to the 'center' of the green, the center of the red, and at 0.8 μm for black-and-white infrared. The situation is slightly different in the case of color infrared films; however we see a peak at 0.55 μm for the yellow sensitized layer, a peak for the magenta layer (red-purple) towards 0.65 μm and a very rounded peak towards 0.7–0.75 μm for the blue-green sensitized layer (cyan).

I have chosen to reproduce here a table compiled by J.-L. Kaltenbach (1970, p. 8). We can see from this table that infrared comes out slightly ahead, followed by red then green. In a more complete table, including for example detection of water pollution, soil erosion, etc., the infrared band would outdistance all other categories of the spectrum (Table III).

TABLE III

Bands of the spectrum recommended for detection of various natural phenomena (J.-L. Kaltenbach, 1970)

Natural phenomena	Blue	Green	Red	Infrared
Vegetation (presence, density, etc.)	/	/	*	*
Distinction between coniferous and deciduous	/	/	/	*
Identification of plant varieties	/	*	*	*
Detection of sickly vegetation	/	/	/	*
(Possible) identification of causes of plant deterioration	/	*	*	/
Showing up dynamic currents of stream meanders	/	/	/	*
Penetration into water, study of shallows	/	*	*	/
Recognition of details in shady regions (from ground only, not from air or satellite)	*	/	/	/
Geological identification	* (on ground)	*	*	*
Urban agglomeration (various aspects)	/	*	*	*

* possible; / impossible

I have borrowed from Gurk (1970) Table IV, which is very significant. I have to point out first that the spectral bands used on Apollo 9 were as follows: 0.51–0.89 μm on infrared color film; 0.47–0.61 μm on panchromatic black-and-white film; 0.59–0.715 μm on panchromatic black-and-white film, and 0.68–0.89 μm on infrared black-and-white film.

Table IV needs no comment, since the importance of color infrared and comparative examination of several bands is clearly brought out. I would like to emphasize, however, the relative mediocrity of the results obtained with black-and-white infrared when it is alone analyzed.

For quite some time, aerial photography has been a trump card in the hands of

TABLE IV

Relative possibilities for interpreting photographs taken by Apollo 9 astronauts (Gurk, 1970)

Plate No. 3799	IR color	Multispectral					
		Individual bands			Analysis of combined bands		
		Green	IR (B and W)	Red	Green and IR	Red and green	Red and IR
Utilization of ground:							
Details	4	4	4	4	4	4	3
Overall	**1**	2	2	**1**	2	**1**	**1**
Ground coverage:							
Types of crops	3	4	4	3	3	3	3
Condition of crops	**1**	4	2	4	2	4	2
Flooded crops	**1**	4	**1**	3	**1**	3	**1**
Pastures	2	2	3	2	2	2	2
Brush	**1**	**1**	2	**1**	**1**	**1**	**1**
Wasteland	**1**	3	3	2	3	2	2
Denuded rocks	**1**	**1**	2	**1**	**1**	**1**	**1**
Dunes	**1**	**1**	**1**	**1**	**1**	**1**	**1**
Totals of '**1**'	7	3	2	4	4	4	5

Note: 1: always interpretable; 2: interpretable with exceptions; 3: rarely interpretable; 4: never interpretable.

Earth scientists. The renewed interest in multispectral photography means that full and outstanding use of this trump will be made, which, after a short digression, justifies the following paragraph.

Photographs taken in three or four different spectral bands must always be taken simultaneously, the three (or four) objectives being focused on exactly the same scene. If possible, the optical axes should be parallel and the cameras should be placed side by side; in this way, the multispectral examination will be happily complemented by a 'relief' study, which will show up several features appearing somewhat attenuated on the individual plates.*

The examination of plates (transparencies) will gain considerably from being carried out as suggested in Figure 4 (p. 17): a backlighted screen and side-by-side projectors project superimposed images through color filters (green, red, and blue). We thus obtain a color reproduction ('composite colors') of very high quality. Of course, as the intensity of the projection lamps must be balanced by filters placed over the apertures, neutral compensating filters must be used rather than rheostats acting on the spectral quality of the projector lamps. It goes without saying that the infrared color transparency must be projected separately.

It might be useful to recall two photographic techniques, one linked to multispectral photography (false colors) and the other one, also in false colors but with only one photograph to be 'enhanced'.

(1) *False colors for the three black and white photographs belonging to the multispectral technique.* Expose successively, a color photographic paper, with the three different black and white negatives, each one with a different filter, namely, red, green and blue. Preferably, use the red filter for the infrared negative.

* The following suggestion may be helpful for the beginner developing exposed films. Ektachrome infrared in a brand new E.4 processing set; black-and-white panchromatics and B & W infrared, all three with the D 19 developer.

(2) *Density splicing* (false colors, for a single photograph). Although very long and tedious, this technique leads to good results, shedding light upon facts completely invisible on the original shot. Three steps are to be followed:

(a) *'Masks'*. The original shot, negative or positive, transferred first to a transparent film will be used to obtain 3 (or 4) contact prints, the first (*M.1*) with normal light intensity; the second (*M.2*) with intensity $\sqrt{2}$ lower than M.1; the third (*M.3*) with intensity $\sqrt{2}$ lower than M.2 etc. The use of an enlarger as a light source is very practical, the enlarger being completely out of focus, even with the smallest diaphragm (aperture). The different diaphragms 1.8; 2; 2.8; 4; 5.6; 8; 11; 16, 22; 32 etc., build up the sequence of decreasing intensity corresponding to our needs (geometric progression with a ratio $\sqrt{2}$). After obtaining the masks, the original photograph is no longer useful.

(b) *'Countermasks'* (CM). Every mask (M.1, then M.2 etc.) originates, by contact, its own 'negative'. But, the same time exposure and the same light intensity must be used, preferably neither the smaller nor the greatest used for the M.1, M.2 etc....

(c) *Obtaining equiplages* (E). On the same transparent film, successively, expose M.1 + CM.2; the next equiplage will be the combination of M.2 + CM.3 etc. Finally we obtain E.1 through E.3 (or E.4). These 'E' documents will serve for the final *Density splicing photograph*. The same virgin color photographic paper is exposed successively with E.1 (red filter); E.2 (green filter); E.3 (yellow filter); E.4 (blue filter). The final result is surprisingly striking, not only by its artistic qualities, but mainly by its scientific value. Long, tedious, but so rewarding.

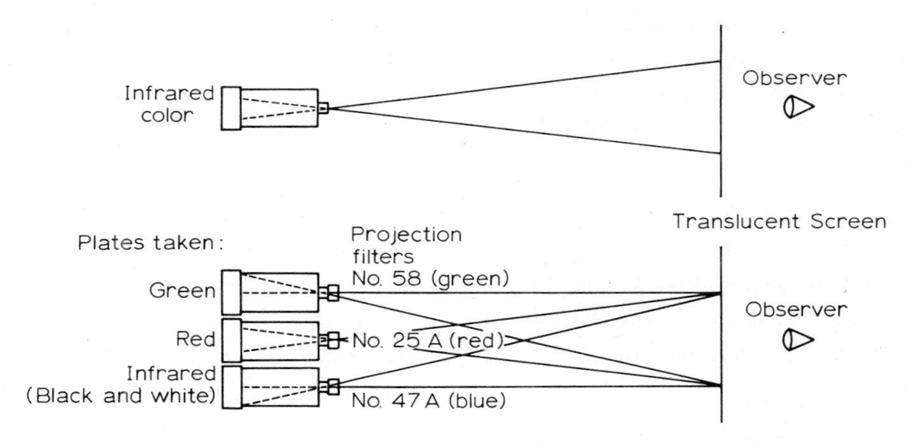

Fig. 4. Reconstitution of a 'composite color' image using three projectors. – The Ektachrome infrared transparancy is examined separately for comparison. Place the 'red' plate, continuously projected, through filter 25A in the center. The 'green' transparency (along with the same filter used in the shot) must coincide, after trial-and-error readjustment, with the 'red' plate. When the left projector is switched off, proceed in the same manner with the right projector (infrared transparency) using a green filter. After the equipment has been set up as described above, use a blue filter (47A preferable to 47) for the apparatus on the right. Rheostats may be used to balance the light intensity, but neutral compensating filters are preferable (choose from 0.1; 0.2; 0.3; 0.6; and 0.9).

3. Stereoscopic Photographs

The photographs taken by the Gemini and Apollo astronauts have shown that, despite the very high altitude and a camera system not permitting the susual hyper-stereoscopic photography, three-dimensional vision was possible and satellites could transmit to Earth pictures that could more easily be processed using past experience acquired in making vertical aerial shots. Accordingly, a brief explanation of the working of 3-D photographs follows:

3.1. Scale

Let F be the focal distance of the camera, A the altitude (or distance) of the camera above the scene, P the distance pp' on the contact print (not an enlargement) and S the same distance on the ground. We can write (similar triangles):

$$\frac{A}{F}=\frac{S}{P}$$

whence $S=PA/F$, gives an idea of the scale, being the quotient A/F.

3.2. Measurements of height differences (see appendix G)

(a) *Utilization of shadows.* The scale has now been established by the above equation. We measure the length of the shadow cast by the object under discussion, and deduce its actual length on the ground. The height of this object is then equal to the length of the shadow multiplied by the tangent of the angle of the Sun above the horizon. It is evident that, to know the angle of the Sun, we need to know the exact date, time, and latitude-longitude position of the shot; all of this data may easily be found in tables. I point this out because, in satellite photography, when the angle of the shot has a spatial resolution *less* than 0.1 milliradian, distances on the ground equivalent to the width of a road, the shadow of a tree, or cliff may be calculated.

(b) *Utilization of the parallax bar.* This procedure is too well known to spend much time on it, so only brief directions are given here – the 'directions for use.'

Let H be the height of the camera (satellite, aircraft, or distance on the ground between the lens and the scene), b the base line on the stereoscopic couple, i.e. the distance between the optical centers of the right-hand and the left-hand photos, and dp the parallax difference. The height of the object studied is thus:

$$\frac{H\cdot dp}{b+dp}.$$

Very often, H must be replaced by a calculated value. To do this: let D be a known distance on the ground (for example between two crossroads), f the focal length of the camera and d the value D on the photographic plate obtained by contact printing, not by enlargement. $H=Df/d$, to be substituted for H in the preceding equation.

It is very important, for common sense reasons, to work on a photographic print obtained by contact from the original negative. It is equally evident that, if we know precisely the value of the linear enlargement, we can replace f by a number obtained when f is multiplied by the enlargement value.

4. Fineness of Resolution of Photographic Emulsions

4.1. Spatial resolution

We know the expressions *fine grain* and *coarse grain*, but the public is less familiar with the notion of resolution – spatial or on the ground. Spatial resolution, or power of separation, means the number of lines per millimeter that a photographic emulsion can *separate* and, thus, the size of the smallest objects that can be brought out on the photograph. For ground shots, and even aerial shots, this has no practical importance, but it is a very different problem where satellites are involved. A good color emulsion has a resolution of 100 lines per millimeter: i.e., under the microscope, one could count 100 white lines and 100 black lines between two lines one millimeter apart.

4.2. Resolution on the Ground

Resolution on the ground is the result of dividing the scale E denominator by the resolution R expressed in lines/mm. Take, for example, a scale of 1:2500000 and a spatial resolution of 100 lines/mm: the resolution on the ground is 2500000/100= =25 m (25000 mm), which is the case for many color photographs taken from manned satellites. The black-and-white emulsions can sometimes give spatial resolutions of 1000 lines/mm which, according to the preceding scale, would enable one to identify an object on the ground with a minimum linear dimension of 2.5 m.

It is worthwhile to stress the merits of remote sensing performed with the aid of photographic emulsions. However, the restrictions are just as obvious, and not because only satellites that are returning to earth can use this technique. We can actually recover films from unmanned satellites – a costly procedure, but it has been done with types of satellites other than those in current use. The essential restriction determines the width of the spectral band within which photographic emulsions can operate.

Looking at the extent of the electromagnetic spectrum – from a few Hz to 10^{-15} Hz (Table I), tempts us to want to explore beyond the microscopic fraction of the continuum within the reach of our cameras.

Since the advent of the space age, marked progress has been made in the frequencies between 3×10^5 GHz (1 μm) and 10^4 GHz (30 μm): the near and middle infrared. For the time being at least, the far infrared is unusable. Research on the 300 MHz (1 m) to 100 GHz (3 mm) frequencies has already produced tangible results despite the small size of satellite-borne antennas. As an example, Nimbus 5 carries a passive radiometer operating in the 19.35 GHz (1.55 cm) band, but resolution on the ground does not meet the needs of earth science specialists, except oceanographers and those interested in the Arctic or Antarctic icecaps.

CHAPTER III

THE INVISIBLE PART OF THE ELECTROMAGNETIC SPECTRUM

(Infrared and Microwaves)

The ultraviolet was recognized in 1801 by J. W. Ritter, who observed the chemical activity of this shortwave radiation. In the previous year, Sir Frederich William Herschel, breaking down light with a prism, noticed that next to the red was a zone where the thermometer showed up hitherto unsuspected energy.

Over sixty years passed before the principles of infrared spectroscopy were established by R. Bunsen and G. Kirchhoff, whose momentous work opened the doors to future research.*

Photographic film is incapable of recording radiations with wavelengths greater than 900–950 mμ. Other instruments then became necessary – radiometers and spectrometers – used also in the visible and the ultraviolet.

Solar energy reaches us after traversing the atmosphere whose density, expressed in grams per cm^3, goes from 10^{-15} at altitude 500 km to 10^{-13} at 250 km, then 10^{-11} at 125 km, 10^{-6} at 50 km, 10^{-4} at 20 km, and finally a little over 10^{-2} at sea level. This increase in density means, among other things, a concentration of various gases that absorb more or less energy from the Sun or from the Earth, sea, and clouds.

1. Atmospheric Absorption

The processes affecting solar radiation or radiation emitted by the Earth's surface are still quite poorly understood. Let us take radiation emitted by the Sun. When solar radiation strikes the atmosphere, part of it is reflected out into space, while the rest penetrates into the atmosphere and is refracted; the refracted portion is, in turn, either sent out into space or in towards Earth. Moreover, multiple gases absorb, in a highly irregular manner, radiations of certain wavelengths. Of the radiation that finally reaches Earth, some types are refracted once again, others are absorbed, while yet a second source of radiation is supplied by the clouds and sky. From this (grossly oversimplified) summary we can see the complexity of the situation. We may know some of the processes, but we must admit that most of them escape us.

Radiation of certain wavelengths is entirely absorbed by the atmosphere, which becomes almost opaque below 0.3 μm and above 15 μm (up to about 1 mm).

1.1. Principal absorption bands

Water vapor, carbon dioxide, and ozone are the chief agents responsible for absorp-

* Kirchhoff established the principles and constructed the first spectroscope. In 1860, Bunsen used the instrument, destined to become a tool of prime importance.

tion, especially in the infrared bands. Nitric acid, methane, molecular oxygen, and carbon monoxide 'share' the less important absorption bands.

(a) *Spectral bands under 400 mμ.* Nitrogen and molecular oxygen make the atmosphere almost entirely opaque below 220 mμ, while ozone does the same from 220 to 300 mμ. Moreover, nitrogen and oxygen prevent any transmission between 30 and 130 mμ. From 130 to 220 mμ atmospheric absorption is practically complete, due to O_2.

(b) *Infrared bands.* Oxygen is an almost negligible nuisance, and then only in the 1.06 μm and 1.27 μm bands, but the relative opacity of the atmosphere is caused by *water vapor, carbon dioxide, ozone*, nitric acid and methane.

(1) Water vapor: The maximum absorption bands are located at: 1.4 μm; 1.9 μm; 2.7 μm; 6.3 μm; and 25 to 100 μm. The minimum absorption bands are observed at 0.9 μm and 1.1 μm.

(2) CO_2: The absorption peaks are located at 2.7 μm, 4.3 μm, and 15 μm, while the minimum is at 2.0 μm.

(3) O_3: Outside the ultraviolet, the main absorption band is centered on 9.6 μm.

(c) *Microwaves.* In the millicentimetric wave region, the atmosphere once more becomes particularly clear. Here we find, chiefly, a water vapor band at 1.35 cm plus two regions for oxygen: 2.5 mm and 5 mm to 1.34 cm.

1.2. Scattering

We have seen the extent to which radiation scattering affects the short wavelengths in the visible, and we know that the phenomenon is governed by Rayleigh's law: scattering is inversely proportional to the fourth power of the wavelength ($1/\lambda^4$). In fact, the rule applies only when the diameter ∅ of the particles causing the scattering is distinctly less than the wavelength. For other relationships between ∅ and λ, Mie's theory applies, roughly, as follows:

$\varnothing < \lambda$, scattering proportional to $1/\lambda^4$;
$\varnothing = \lambda$, scattering proportional to $1/\lambda^2$;
$\varnothing = 3/2\lambda$, scattering proportional to $1/\lambda$;

and finally, when $\varnothing \geqslant 2\lambda$, scattering is independent of the wavelength.

For example, particles of fog with diameter approximately 250 mμ cause accentuated scattering in the ultraviolet, violet, and blue; when the droplets reach greater dimensions, such as 8 μm, the mean diameter of fog particles, scattering affects the near and middle infrared.

2. Atmospheric Windows

Figure 5 shows the main atmospheric absorption phenomena from 1 μm to 15 μm.

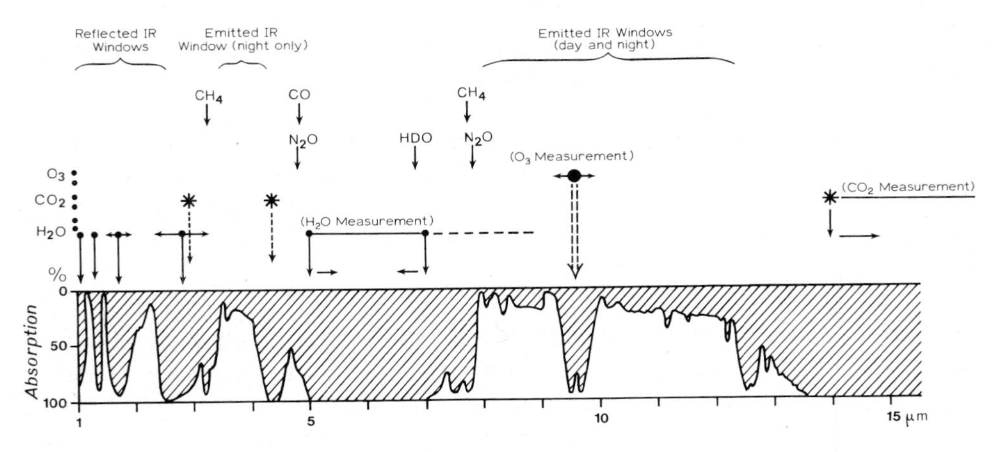

Fig. 5. Atmospheric Absorption. – The light-colored zones are the bands with least absorption (atmospheric windows).

The shaded zones are more extensive than the white zones; the latter represent bands of the spectrum that let through radiation. These white zones are known as *atmospheric windows*, and allow emitted, re-emitted, and reflected radiation to pass through.

2.1. Distribution of Atmospheric Windows

We have four atmospheric windows, namely:

(a) *0.3 to 1.3 μm.* This window includes all of the visible and part of the near ultraviolet and the near infrared. It is used by photography (up to 0.9 μm), television (up to 0.9 μm) spectrometers, and radiometers. In every case, detectors pick up the reflected radiation.

(b) *1.3–2.5 μm window.* This spectral band, with the end section of the above band (0.9 to 1.3 μm) is beyond the range of photographic emulsions and television cameras. Only spectrometers and radiometers can operate in this region, picking up only reflected radiation. The infrared energy emitted is much too weak to be detected. This would not be the case if the temperature of the objects could be increased by several thousand degrees centigrade.

(c) *3.5–4.2 μm window.* Here we are entering the domain of emitted infrared whose energy is sufficient to be picked up by radiometers and spectrometers. This window was used on the 1964 and 1966 satellites (Nimbus). Unfortunately it only works at night, since, in the daytime, emitted and reflected radiations combine and they are difficult, though not quite impossible, to separate.

(d) *8–14 μm window.* If measurements are made from an altitude above the ozone layer, only the bands located to either side of 9.6 μm are valid. This window is the most important of all for it works both day and night; reflected radiation is negligible. Moreover, it is precisely at this wavelength that maximum ground emission is observed, when the temperature is around 300 K, at ±30 K (300 K=27 °C, and 1 K=1 °C).

The opacity of the atmosphere prevents working in bands higher than 14 μm. On the other hand, as soon as we get into the microwave range, the atmosphere becomes transparent again. A special paragraph will be devoted to these important microwaves.

2.2. The Local Minimum, or Dip (Reststrahlen)

The continuous curve of radiations plotted against wavelengths shows two *dips*, one between 8 and 11 μm, the other from 18 to 25 μm, the latter being noted here only for completeness as it affects a spectral region of no interest to earth science. The silica-oxygen rotation is responsible for the phenomenon, so we see immediately that the 'depth' and location of these dips are governed by the silicate content of emitting objects such as soils and rocks. When the rocks are basic the minimum appears at a longer wavelength than when they are acid. Figure 6, after Hovis (1968) illustrates the fact:

(a) *Basaltic rocks* (curves 1 and 2 of Figure 6). From 8 to 9 μm, the *radiance temperature* (equivalent blackbody temperature) is less than that recorded at 11–12 μm. This rock was chosen by the experimenter since "... it possesses the weakest *Reststrahlen* of most igneous rocks" (Hovis, 1968, p. 29). It would be difficult to say whether the dip observed at 9.7 μm is a real one or is merely due to phenomena peculiar to the soil. Moreover, this dip would not be recorded from a satellite since it is located just inside the O_3 absorption band.

(b) *Sandy (siliceous) rocks*. In this case, the dip is strongly accentuated towards 8.5 μm. The heat contrast between the 8–9 μm and 10–11 μm spectral bands is just as pronounced. The fact has momentous importance for earth science since it enables

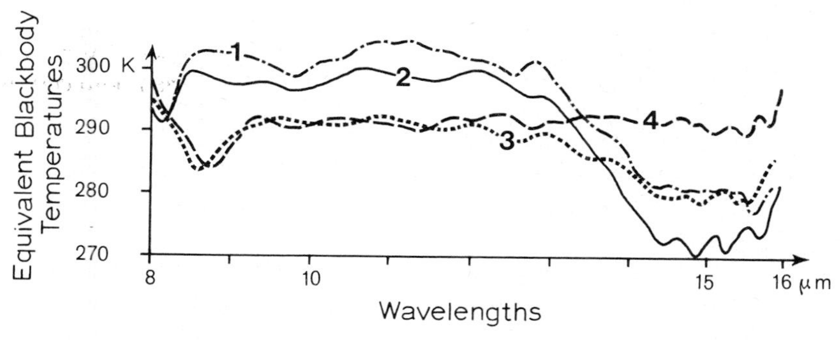

Fig. 6. Pattern of radiance temperatures (T_{BB}) between 8 and 16 μm. – 1 and 2: volcanic flow at 17:22 and 17:46 GMT; 3 and 4: dunes of gypsum sand at 15:42 and 16:28 GMT. (After Hovis, 1968. Region of El Paso, Texas.)

us to acquire better understanding of natural phenomena (distinction between acid and basic rocks). This singularity has been made use of only for isolated pieces of research. The Nimbus 5 satellite (1972) was the first to carry a radiometer with two channels, on either side of the dip we are discussing (but it went out of order almost immediately).

2.3. Utilization of atmospheric windows

It is hardly necessary to stress that the spectral bands centered on 6.3 μm, 9.6 μm, and 15 μm are of enormous importance to meteorological sciences, since they permit measurement of atmospheric water vapor, ozone, and carbon dioxide; the latter leads to information on the thermal values of the stratosphere. However, earth science research requires that emitted or reflected radiation be let through, and we are thus left with only the following windows:

(a) *Visible and near infrared*, exclusively reflected radiations from 0.4 to 1.3 and 2.5 μm.

(b) *Middle infrared*, 3.5 to 4.2 μm. Usable only at night.

(c) *Middle infrared*, from 8 to 14 μm. Usable day and night.

Windows (b) and (c) above supply, after calculation of the radiation received, radiance temperatures different from their real temperatures. As we will have occasion to see, the real temperatures have considerable effect on the radiance temperatures, which we will often term 'equivalent blackbody temperatures,' T_{BB}.

The influence of the real temperature, predominant during the night, gives way to that of emissivity during the day. This will be discussed later in more detail.

3. Microwaves

This spectral region, well known to the public because of radar, commences at around 1 mm and extends to the meter waves. It has both advantages and disadvantages, the latter with respect to the position in which an antenna must be placed. This applies to both passive and active microwaves.

3.1. Active microwaves (radar)

In about 1920, based on the observation that radio waves fluctuated due to passage of aircraft, the idea of receiving a wave that had been emitted then bounced back by an obstacle gave birth to successively refined apparatus that finally became radar in 1939. The word is the result of the somewhat whimsical contraction of *R*adio *D*etection *A*nd *R*anging. The 'detection' is the disclosure of the object by the reflected signal, and the 'ranging' the measurement of its distance, by a change in frequency of the return signal.

Radar is probably unique in the world of remote detection, since it possesses its own source of illumination (emitted part). The relative distances and speeds are calculated using the principle of the Doppler effect, i.e., the change in frequency of the return signal as modified by distance. This principle is becoming well known to drivers who overshoot the speed limit.

For security reasons in the Second World War, radar bands were designated by letters; the custom was retained even after the need for security had disappeared. At present, this letter system is not uniform among manufacturers, but Table V gives the most widely used distribution. The frequencies universally employed are shown.

TABLE V
Radar bands

Band		Wavelength (in cm)	Frequency (in GHz)
P		136–77	0.22–0.39
L		77–19	0.39–1.55
S		19–5.76	1.55–5.2
	A (Phillips)	10	3
C		7.69–4.83	3.9–6.2
	J	5	6
	H	3.75	8
X		5.76–2.75	5.2–10.9
	J, or Ku	2	15
K		2.75–0.83	10.9–36
	Q, or Ka	0.78	38
	Y	0.6	50
	O, or E (Phillips)	0.4	75

(a) *The elements of radar.* The principle of radar is fairly simple: a radio wave is beamed at an object which, depending on its own characteristics, reflects the incident wave whose 'qualities' (intensity, frequency, polarization) can be analyzed. The back radiation depends on dielectric and conducting properties, roughness, slope, etc., of the reflecting surface, so that one surface can be distinguished from another by its different properties. For example, vegetation is an excellent reflector due to its water content; withering plants reflect less, etc.

(b) *Angle of incidence.* Perpendicular beaming onto the target is avoided since the vertical (angle of incidence 0°) complicates distance measurements for geometric reasons. The angle of incidence is thus somewhere between 0° and 90°, both of these extremes being avoided. When radar is beamed from an aircraft, the angle of incidence is between 20°–30° and 75°–80°: this is usually called *side-looking radar*. The greater the altitude of the airborne platform, the more difficult it is to avoid the vertical.

(c) *Polarization.* Polarization is determined by the design of the antenna (electric field vector). The reception antenna can be designed to respond only to the original polarization, or to receive other polarization. It is possible to use several different receivers, each for one kind of polarization; vertical, horizontal and 'circular'; in such a case, shifts in polarization can be observed, which contribute to accurate identification of the objects being studied.

We can see how extensive the possibilities offered by radar *would be* for earth sciences. Unfortunately, I have had to emphasize the restrictions when considering a carrier platform such as an aircraft; these restrictions could continue far into the future for both manned and unmanned satellites.

(d) *Atmospheric interference*. The difficulties of using radar, as distinct from infrared and the visible, do not stem from atmospheric obstacles. Absorption by O_2 and water vapor is negligible. Clouds are 'traversed' since the *echo* from a drop of water is proportional to $\varnothing^6/\lambda^4$ ($\varnothing$: diameter of droplet, λ: wavelength); thus clouds do not echo unless they are rainclouds (large droplet diameter) and the wavelength is sufficiently short. This feature is exploited every day by weather bureaus to detect rainfall areas.

Absorption by water droplets is inversely proportional to the wavelength and the cube of the particle diameter ($1/\lambda \cdot \varnothing^3$): as a result, total absorption is governed not by the distribution of the individual diameters but by the density of the mass of water drops. Experience has shown that abundant rainfall may disturb radar operation when the wavelength is in the vicinity of 1 cm, but has no effect when the wavelength is greater than about 8 cm.

(e) *Antenna problem*. The size of the antenna governs, with respect to the wavelength, the spatial and consequently the ground resolution. Table VI gives only a

TABLE VI

Antenna size required for microwaves
(after Holter, see Johnson (P. L.) 1969)

Antenna diameter (m)	Frequency (GHz)	Wavelength (cm)
12 to 14	5	6
10	10	3
8 to 9	15	2
7 to 8	20	1.5
6 to 7	25	1.2
Near to 6	30	1

rough approximation, leading to a spatial resolution of 5.1 mrad (top line) to 2 mrad (last line). From an altitude of 1000 km, this would lead to linear ground resolution of 5.1 to 2 km. It would be difficult to mount such antennas on a satellite where they could obstruct other scientific equipment.

Let us recall that the linear ground resolution is a function of the spatial resolution and distance (altitude of the satellite, or aircraft) from the scene. In the microwave region, the *spatial resolution*, i.e. the antenna beamwidth, is equal to 1.2 (λ/D) where λ=wavelength and D the antenna diameter, with the result expressed in radians. The linear ground resolution is the product of the spatial resolution and the distance from the scane (heigh, H). Finally, the following equation gives this ground resolution:

$$\frac{1.2 \cdot \lambda \cdot H}{D}.$$

These observations must be kept in mind when passive microwaves are considered. For the earth sciences, the ground resolution is of vital importance. However, it is the author's feeling that one should seek neither too big nor too small a linear ground resolution. Ten kilometers is the extreme limit, but 1 km should also be the lower limit; better still, between 500 m and 1000 m. Resolution finer than 500 m is the province of the aircraft, at lower cost.

From a spacecraft like ERTS, with the very useful wavelength of 1.55 cm (19.35 GHz), the antenna diameter should be around 20 m to obtain a 930 m ground resolution from 1000 km. I do not believe that such a payload should be placed aboard any ERTS, but the passive microwaves rightfully belong on the Spacelab built jointly by NASA and ESRO for 1979–1980. As the orbital altitude is about 500 km, a 20 m diameter antenna, easy to install, would give a linear ground resolution of 465 m, far more than actually needed. For the record only, a NASA representative, at a post-Apollo symposium held by ESRO in February, 1973 at Frascati, Italy, proposed to launch at 100 m diameter antenna, with Spacelab, to give a linear ground resolution of 93 m.

(f) *Choice of wavelength and advantages of radar*. The span of available frequencies is vast but it seems that experts have not yet made their choice as to the most suitable wavelengths for the various aspects of earth sciences research. However, a study of water systems on denuded or wooded land would do well to use decimeter and meter wavelengths, also ideal for soil study. Vegetation types could be distinguished easily by using higher frequencies in the centimeter bands. The advantages of radar appear well established for every field of earth science (geology, hydrography, botany...) but, we repeat, these advantages disappear with satellites and increase with aircraft.

3.2. Passive microwaves (naturally emitted by objects)

The future prospects for passive receivers in the centimillimetric bands appear more promising than for radar. This opinion is based on two facts: (1) the power source necessary for radar does not exist; (2) the radiation naturally emitted by the objects is a better reflection of the 'personality' of these objects.

These are thermal radiations. According to Planck's law, electromagnetic radiation is a function of the temperature of the object and the wavelength. In the microwave domain, an approximation of this formula gives:

$$W_{BB} = T_{BB}^{-4},$$

where W = radiation, T = temperature and BB = blackbody. This applies where a blackbody is taken as a reference. The brightness temperature T_{B_r} of real objects, o, is a function of emissivity, ε, at the wavelength considered, λ, namely:

$$\varepsilon = \frac{W_o \lambda}{W_{BB}}$$

T_{Br} derives from these basic concepts.

The radiation emitted and picked up by the antenna depends on three factors: emission, reflection, and transmission. The energy emitted is proportional to $\varepsilon\lambda$ and to T_o (real temperature of object) and is affected by the spectral reflectance, $\varrho\lambda$, the transmittance, $\tau\lambda$, and finally by the consequences of radiations incident on the object, T_s coming from the sky and T_c coming from the clouds, which gives, for the value of the brightness:

$$T_{Br} = (\varepsilon\lambda \times T_o) + \varrho\lambda T_s + \tau\lambda T_c .$$

In fact, the last two terms of this sum can be neglected without affecting the value T_{Br} too much. It is evident that in the case of a blackbody where $\varrho\lambda = \tau\lambda = 0$ and where $\varepsilon\lambda = 1$, $T_{Br} = T_{BB}$, and, in the case of actual objects, that T_{Br} is thus the product of the real temperature (expressed in energy units) and $\varepsilon\lambda$. As an absolute value, this formula is incorrect but, as a relative value, i.e., compared with related facts, this simplified form is very acceptable.

All of the above also applies to thermal infrared. In this domain, $\varepsilon\lambda$ varies rather little such that T_{BB} derived from the effective radiance is fairly near to T_o, but never equal to it, and the differences $T_o - T_{BB}$ vary according to the objects considered, and according to the time of measurement (night or day). At centi-millimetric wavelengths, the variations in emissivity are considerable.

All these facts suggest an idea already proposed to the French CNES and the European ESRO. Let me first emphasize the fact that, in the thermal infrared, the changes of emissivity are rather weak while, in the passive microwaves, they are strong. For example, pollutants added to water provoke only a few degrees of difference in the infrared, but much more in the microwaves. On the other hand, the T_{BB} are rather close to the real temperatures, while the T_{Br} are subject to the tremendous changes in emissivity. Consequently, the ratio between T_{Br} and T_{BB} (in energy units) gives a fair approximation of the emissivity value in the milli-centimetric wavelengths.

The suggestion is simple. Associating an infrared and a passive microwaves radiometer is the simplest method of obtaining the emissivity value. Naturally, both radiometers should have the same ground resolution and should 'look' at the same scene. There is no problem for the ground resolution, but the second facet of the problem is a tricky one (geometry); however, I have complete faith in the skill of our specialists who have already performed miracles. I would suggest a spatial resolution of 2 mrad (antenna, for the passive microwaves, having a diameter of 9.3 m, the wavelength being 1.55 cm). Such a spatial resolution, from an altitude of 500 km, would give a ground resolution of 1 km. Naturally, with only 1 milliradian (antenna diameter 18.6 m), one would observe terrestrial features of 500 m.

Operating under cloudy conditions, only the microwave data would 'give' the terrestrial features. Under cloudless conditions, infrared and hyperfrequencies would operate simultaneously, and in addition to the two images, we would obtain a third, that is the emissivity pattern, giving a more accurate 'description' of the personality of the objects we were looking at. Such a new document, either as a photofacsimile, or, better, as a computer-produced grid print map – digitized data – would be positively priceless.

Such a payload, let us hope, will be a characteristic of Spacelab, thinking of remote sensing of Earth resources.

This chapter has enabled us to touch upon some technical aspects, knowledge of which is essential for interpreting information gathered at a distance. Multispectral photography preferably associated with three-dimensional techniques, and exploitation of the infrared and microwave bands, open still-unsuspected prospects to the earth sciences, whether the carrier platforms are on Earth or are air- or satellite-borne. In the next chapter, we will drop back to Earth and look at some of these future prospects.

CHAPTER IV

EMITTED RADIATION AND REFLECTED RADIATION: CONCRETE ASPECTS

We have already looked at the central topics in regard to centi-millimetric waves; we now need only clarify some ideas on the near and middle infrared, which is now being exploited intensively.

1. Reflected Radiation *

1.1. Comparative values

Only Nimbus 3, 4, 5, and the ERTS have used or are using the reflected near infrared in the 0.7–1.3 μm band, while the response is effective only beginning at 0.8 μm. Some research has already been done in this respect, with many different ends in view. First of all, ecological concern has motivated scientists to make observations on the reflectivity of various types of vegetation. Next, in preparation for geological reconnaissance of the Moon and the outer planets, some bases for interpretation have been investigated. The result is a happy one for earth science, since we have available some observations, although fragmentary and highly inadequate.

In 1960, G. H. Suits** emphasized one important phenomenon:

The reflectivity power (of solar energy) decreases as the wavelength increases, until the radiation emitted by the object is dominant. *The point of intersection where emitted radiation is greater than reflected radiation is at approximately three microns.*

(a) *Rocks.* This 3 μm limit is the one we will adopt for the transition between reflection (below 3 μm) and emission (above 3 μm), which by no means signifies that one cannot measure reflected energy over 3 μm: some investigations have had the precise aim of studying the pattern of reflectivity up to 25 μm, for example those of Hovis and Callahan (1968). Their research had the purpose of assisting in geological reconnaissance of the Moon and the outer planets. As an example, here are some results obtained by these authors; they are presented not in their original form (curves) but as a table (Table VII).

We can see immediately that the highest values and sharpest contrasts are found at wavelengths under 3 μm, with the exception of the basalts. On the other hand, size

* Numerous publications have been devoted to reflectivity, especially in the visible. Although it is now quite old, this work deserves mention: E. L. Krinov (Laboratory of aeronautical methods, U.S.S.R. Academy of Sciences), *Spectral Reflectance Properties of Natural Formations*, National Research Council of Canada, Technical Translation TT-439 (translated by G. Belkov), Ottawa, 1953.

** Suits, G. H. (1960), 'The Nature of Infrared Radiation and Ways to Photograph it', *Photogr. Engin.* **26**, 763–772.

differences in the materials cause striking contrast below this limiting value where reflectance rises steeply with diminishing particle size.

Beyond this (longer wavelengths), the contrasts are less sharp, the absolute values are smaller, and intersections between curves can cause some confusion in interpretation.

TABLE VII

Reflectivity as a function of size of rock materials
(Values of curve peaks. Reflectance as a percentage. Local minimums: --------)

Wavelengths	Intact rock (%)	Fragments of 105–205 μm (%)	Fragments <38 μm (%)	Local minimums	Rock types
1 μm	37	45	55		
2 μm	40	67	69		
--------	--------	--------	--------	2.8 μm	
3.5–4 μm	20	50	65		
--------	--------	--------	--------	8.9 μm	GRANITE
9 μm	21	15	9		
15 μm	4	2	4		
22 μm	15	9	6		
1 and 2 μm	5	14	40		
--------	--------	--------	--------	2.8 μm	
4 μm	5	9	30		
--------	--------	--------	--------	8.9 μm	SERPENTINE
10 μm	24	7	4		
15–16 μm	21	5	3		
21–22 μm	38	9	5		
0.9 μm	4 (31)	8 (72)	15 (82)		
2.6 μm	4 (36)	7 (82)	20 (93)		
--------	--------	--------	--------	2.8 μm	
4 μm	5 (low)	9 (low)	24 (low)		
--------	--------	--------	--------	4.9 μm ?	BASALT AND
9.5 μm	15 (49)	6 (29)	4 (12)		(QUARTZ)
15 μm	4 (4)	1 (1)	1 (4)		
20 μm	8 (30)	4 (19)	4 (8)		

(b) *Vegetation.* Figure 7 shows the usefulness of reflected infrared for differentiating types of vegetation according to their foliage, and the state of health of a given type of plant. We see the enormous value of the 0.8–1.3 μm since the absolute values are clearer and the contrasts sharper. The dip at 1.4 μm can also be an excellent means of identifying plant species or types of rocks. In this respect, too few systematic studies are available.

Chlorophyll has an important part to play from 0.4 to 2.6 μm. In the visible, it dominates the spectral region due to two very strong absorption bands at 0.45 and 0.65 μm. In the near infrared, with the exception of the 1.45 and 1.95 μm bands which are characterized by H_2O absorption, chlorophyll becomes a good reflector.

Looking at maple leaves and contrasting the green (healthy) with the red (withering)

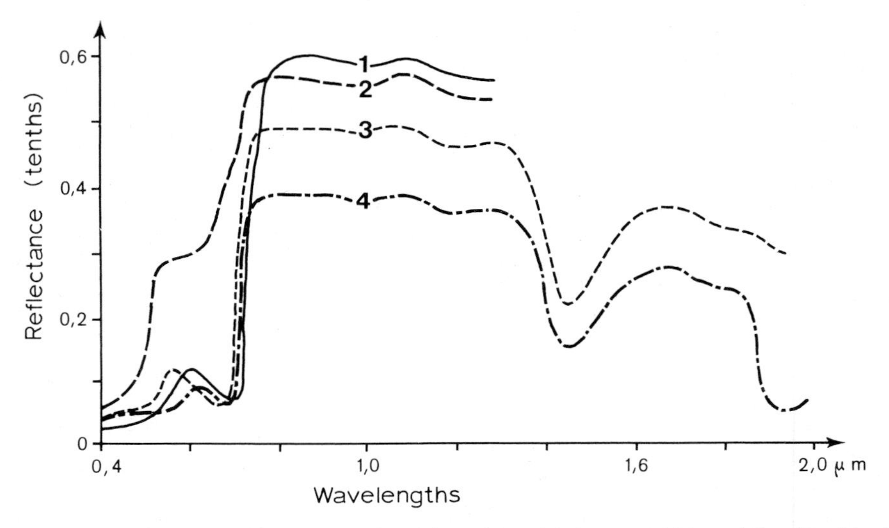

Fig. 7. Pattern of reflectance of 400 mμ at 1.3 and 2 μm. – 1 and 2: leaves of African violets, No. 1 healthy, No. 2 withering (after Knipling, 1969; see Johnson); 3: oak leaves; 4: maple leaves (after Dr McLeod, NASA Goddard Space Flight Center, personal communication).

leaves, we note that the latter have high reflectance below 0.6 μm and above 0.9 μm with a dip, as predicted, at 1.5 μm, with no change shown in the relationships between healthy and dead leaves.

These remarks apply to all plant life, though of course, the absolute values differ from one species to another. For example (see Hoffer in Johnson, 1969), from 0.8 to 1.4 μm, the tulip tree (*Liriodendron tulipifera*) is characterized by reflectance 5% greater than that of the silver maple (*Acer saccharinum*) which has 5% more than the American elm (*Ulmus americana*). At the dip (1.5 μm), the positions are exactly reversed.

1.2. Albedo and reflectance: numerical values

It would be desirable to reduce the photographic data in the visible and near infrared to figures, so that we could deal only with quantitative values. Microdensitometric processes, when applied to the original negative, are of invaluable assistance but, in my opinion, too rarely used. As soon as we have signals picked up by radiometers and spectrometers, we are 'going back to the source' by operating on energy values. Indeed, it is my opinion that one should retain the *effective radiance* expressed in watts per square meter and per steradian; this would avoid much confusion introduced by 'temperatures' or even 'percentages'.

The *albedo* is the amount of solar energy reflected (from a given object). It is calculated for a spectral band extending from 0.2 to 4 μm; this region corresponds to almost all (99%) of the solar energy reaching the ground *. The average resolution of

* Solar energy arrives at the ground in short wavelengths (think of the temperature of the sun); re-emission by objects is at longer wavelengths. Radiation from objects that have absorbed part of the (incident) solar energy usually peaks at about 10–12 μm. At 4 μm, the energy emitted by objects is still very weak, with interference from reflected energy, as has already been stressed.

the infrared radiometer on the Nimbus satellites enables calculations to be made quickly, since the No. 5 channel picks up precisely the energy corresponding to this spectral band.

In the case of smaller spectral bands, usage favors the term *reflectance* or an explanatory formula such as *near-infrared albedo*. In both cases, the following equation always applies:

$$R(\lambda\varphi)=\frac{\pi \cdot N(\lambda\varphi)}{S \cdot \cos\zeta(\lambda\varphi)},$$

where R = the reflectance or albedo;
$\lambda\varphi$ = coordinates of region considered (latitude and longitude);
$\pi \cdot N$ = radiance in the spectral region used;
S = solar irradiance in this spectral region, i.e., 0.7 to 1.3 μm or, in the case of total albedo, 0.2 to 4 μm;
ζ = the zenith distance from the Sun at the time of observation.

Although this formula is easy to employ when the information has been gathered on the ground or from an aircraft, it is not as simply done with instruments on board a satellite that scan vast areas from east to west (subpolar satellites), and all the more so as we get a considerable hour-to-hour time increment where high latitudes are involved. In this case, errors of interpretation can be partially avoided only if we keep in mind both the procedure used to give reflectance or albedo percentages and considerations of the actual time at the regions considered. As a non-isolated example, if we do not consider these two factors, we can wrongly state that in the spectral band 0.7–1.3 μm, cloud reflectance is greater at low than at high latitudes – an error that has already been committed by forgetting about time compensations; as a result, the zenith angle was incorrectly 'manipulated.'

This problem of reflected radiations crops up again with emitted radiations.

2. Emitted Radiation

We already know that we can only record the infrared emitted by objects on the ground through narrow atmospheric windows. As the two windows 3.5–4.2 μm (nighttime only) and 10.5–12.5 μm (day and night) are available, we are confined to these spectral bands, the second being more useful.

We have seen that, by Planck's law, the radiance temperature is a function of the real temperature and the wavelength. We have already noted that all objects, provided their real temperature is greater than absolute zero (0 K = (−273 °C), are affected by molecular agitation, which is translated by an emission of electromagnetic waves whose maximum intensity is located at a wavelength that is a function of the real temperature. As an example, this wavelength is situated at:

2.5 μm, when the real temperature is 1100 K (827 °C)
2.75 μm, when the real temperature is 1000 K (727 °C)

3.6 μm, when the real temperature is 800 K (527 °C)
4.7 μm, when the real temperature is 600 K (327 °C)
5.6 μm, when the real temperature is 500 K (227 °C)
10 μm, when the real temperature is **300** K (**27** °C)

The last line is in boldface since the average temperature of the various objects on the surface of the Earth – water, plants, rocks, soils, etc. – is in the vicinity of 27 °C, and a difference of 20° to 30 °C has very little effect on the 10 μm wavelength. As a result, the window centered on 10–11 μm is the most interesting of all, since it receives the maximum possible energy emitted by the objects we are studying (see Appendix F).

2.1. Calibration

By calibration, I mean the translation into radiance temperatures of signals detected by the instrument, and recorded on magnetic tape. Taking into account the wavelength considered, we can write:

$$\bar{N} = \int_0^\infty B_\lambda(T_{BB})\, \varnothing\lambda,$$

where:
$\bar{N}$ = effective radiance in W/m^2/ster;
B_λ = Planck's constant;
T_{BB} = radiance temperature (equivalent blackbody temperature);
$\varnothing\lambda$ = effective spectral response in the wavelength considered.

In the laboratory, we establish a relation $T_{BB} \leftrightarrow \bar{N}$ and the blackbody used completely fills the field of view of the instrument. More simply, the correlation is established between intensity of radiation emitted by an object and its real temperature. Figure 8 shows the relations $\bar{N} \leftrightarrow T_{BB}$ for the two atmospheric windows.

This calibration was done at the Goddard Space Flight Cnter using a setup similar to that on the satellite. The heat source is supplied by a tungsten-iodine lamp whose radiations, reflected by a concave mirror, pass through spectral filters, neutral filters, a vacuum chamber, and are finally directed onto the radiometer.

Unfortunately this calibration, while satisfactory for temperatures above about 260 K, is more than doubtful below 240 K (Williamson, 1970); interpretation of these low radiation values is extremely ill-advised as errors reach and exceed 20 K.

2.2. Emissivity

Remember that the radiance temperature is that which a blackbody would have when placed under the same conditions as the object studied. It is the product of the real temperature (expressed in energy units) and emissivity. The latter factor is thus of considerable importance. Water, very moist bodies, and those with a very dark color have emissivity approaching 1. All other objects have a ε diverging from 1 and sometimes reaching very low values.

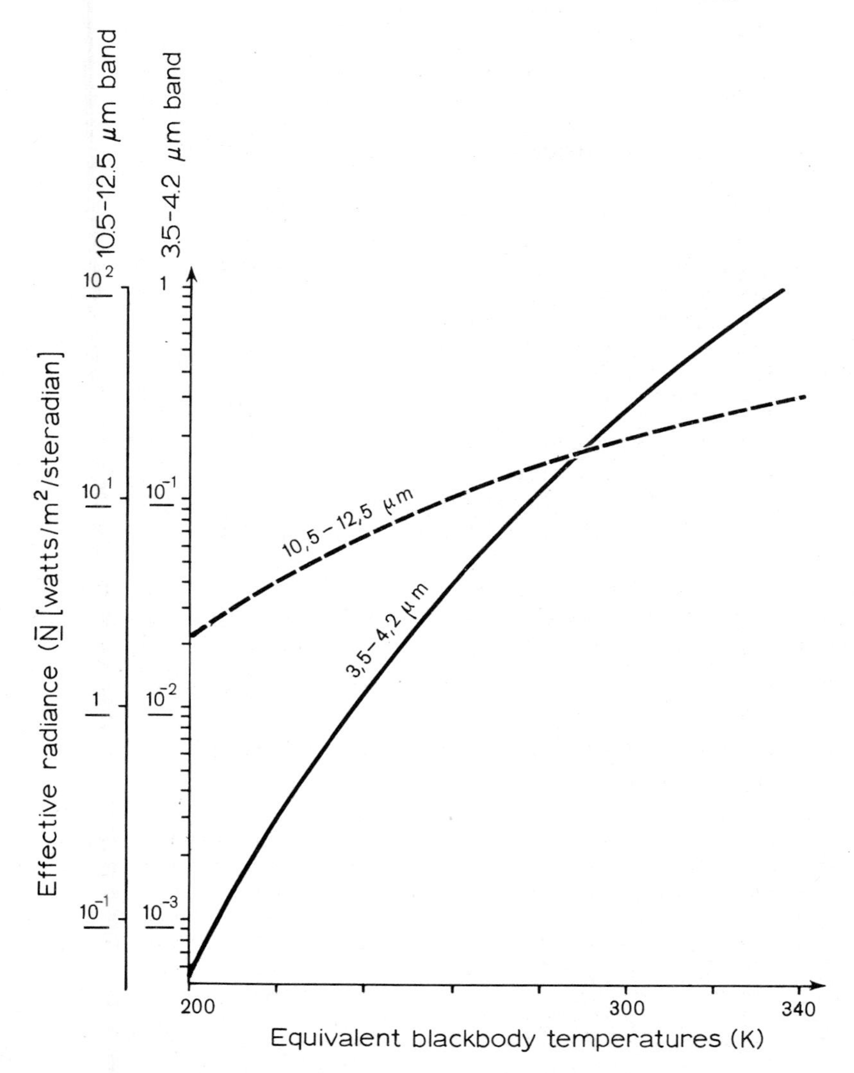

Fig. 8. Effective radiance curves, equivalent blackbody temperatures. (High resolution infrared radiometers, Nimbus 2, Nimbus 4, and Nimbus 5.)

Table VIII furnishes, not emissivity values but the differences between real temperatures measured with a standard thermometer at the very instant when the radiance temperature was recorded by a radiometer operating in the 10.5–12.5 μm band (Pouquet, 1972, *Rev. Géom. Dyn.*).

Buettner (1970) in his communication to the symposium on remote detection at Leningrad, emphasized the role of emissivity. He showed the extent to which discrepancies in actual temperatures could be important according to the spectral band and the objects considered. For example:

A *granite powder* in the *3.8 μm band* has an *emissivity* of 0.37, the temperature

TABLE VIII

Relationship between real temperatures and equivalent blackbody temperatures

Samples measured	Temperatures	
	Blackbody	Real
Basalt, with desert varnish	19.5 °C	20.5 °C
Same basalt, without desert varnish	17 °C	18.5 °C
Coarse granite alluvium (∅ 1 mm)	10 °C	14 °C
Fine granite alluvium (∅ 0.5 mm)	11 °C	16 °C
Running water, clear	13 °C	16 °C
Same water, muddy	14 °C	16 °C
Stagnant water, very salted	11 °C	13 °C
Salt stagnant water rich in microorganisms	9.5 °C	16 °C

The emissivity of the above samples is, respectively, 0.98; 0.97; 0.93; 0.92; 0.95; 0.96; 0.96; and 0.90 (see Appendix F).

increment (ΔT) being 26 K (1 K = 1 °C). In the *11 μm band*, $\varepsilon = 0.93$ and $\Delta T = 5$ K.

Sodium chloride has an *emissivity* of 0.3 and $\Delta T = 30°$ in the 3.8 μm band, these respective values being 'very low' and 'very high' at 11 μm.

Grass prairies have an *emissivity* of 0.76, ΔT of 0.7° at 3.8 μm and ε: 0.97, ΔT: 2° at 11 μm.

Water and ice, very near to the blackbody, have ε: 0.98, ΔT: 0.5° at 3.8 μm, and ε: 0.99, ΔT: 0° at 11 μm, etc.

The state of affairs is a little more complex than this: ΔT, as long as ε is constant, changes with the real temperature. Assuming (wrongly) that ε does not vary with changes in temperature, when, in the spectrum band 10.5–12.5 μm, $\varepsilon = 0.5$, $\Delta T \approx 34°$ at −10 °C, which deviation becomes 44° for $T = 40$ °C and 50° at 80 °C. Taking $\varepsilon = 0.9$, ΔT passes for the same real temperatures as above to, respectively, 6°, 9°, and 10°. Finally, for $\varepsilon = 0.98$, these deviations are 1°, 1.5°, and slightly above 2°. In fact, the following variations grow logarithmically (J. Pouquet, research in progress).

I assumed that ε did not change with thermal variations, but this is not the case. A student of Buettner, R. W. Dana, published an extremely important master's thesis in 1969 on emissivity in the 8–14 μm band. He showed that ε changes with real temperature. For example for rocks such as quartz, granite, gabbro, dunite and albite, and for water: when $T > 0$ °C, $\varepsilon < 1$; when $T = 0$ °C, $\varepsilon = 1$ and finally when $T < 0$°C, $\varepsilon > 1$ (Dana, 1969, p. 40). Quartz, olivine, potassium feldspar, and microcline tend to reduce the ε value while this value increases when darker rock shades are due, among other things, to the presence of augite, labradorite, and hornblende.

Having reached this point, I think advisable to recall what has been said about emissivity in the microwave regions: the variations of ε are considerable in the milli-centimetric wavelengths, the spectral bands of the future, the thermal infrared belonging already to the past.

The role of emissivity, although a logical one, is often underrated. Some investigators are trying at all costs to find the exact temperatures of oceans from the equivalent blackbody temperatures obtained by satellite observations. To do this they go into sophisticated 'correction' calculations adding, as the case may be, 2°, 3°, 5°, or 10 K, and compare their results with those of marine researchers who dip a thermometer into a bucket of seawater (in fact, these 'corrections' are 'calculated' in order to make their results agree with the field results). Even if it were possible to get the exact temperatures, which would assume knowledge of atmospheric processes still not available to us, how could the values of the water surface (infrared has no power of penetration) be compared to those of a bucket of water hauled up from the depths? I have just alluded to the 'surface' and I think it useful to reproduce here a long quotation from Buettner (1970):

For many colleagues... the surface is none other than a botheration. We can compare this surface to the Chinese princess Turandot tangled up in enigmas and obstacles of resolution on the ground, scattering, atmospheric absorption.... Then Grimm's Cinderalla turns up and asks childish questions: what's it for? So what is the temperature of the surface of Paris (the original text refers to Leningrad)? What does it mean?

This question was raised with respect to 'the surface.' It is an excellent one but, with Buettner, we will answer it only in the third part of this book.

CONCLUSION TO PART 1

The first part has shown the vast possibilities offered by the atmospheric windows available on satellites, aircraft, on the ground, and by photography. I would like to conclude by bringing out some remarkable facts.

(1) *Photography*, especially multispectral photography, is undergoing a resurgence of interest. The procedure is simple, inexpensive, and its results, though often discounted, have been neglected for too long.

(2) *Radiometers and spectrometers* should be part of the standard equipment of all our earth science labs.

(3) More and more *satellites* are being launched, and information is pouring in at an ever-accelerating rate. For example, 'our' first satellite, ERTS 1, can provide 300000 (three hundred thousand) documents a week looking like photography (TV channels, spectrometric channels). At present, magnetic tapes are piling up in thousands and hundreds of thousands, waiting for properly-trained investigators (not just 'compilers' who exploit the results obtained by others designated by the contemptuous term 'technician').

(4) Most of the developped countries are launching *air survey programs* for studying the environment. In the United States, the U.S. Geological Survey, State geological services, agricultural services, and private companies are systematically making flyovers in aircraft carrying the same instruments as those on board satellites. This is an obligation from which no large country can abstain.

(5) In a short time, *remote detection techniques* will be *taking the first place in the earth sciences*, without, of course, replacing traditional techniques: the old lab methods and field studies are becoming increasingly imperative.

The exploits of the astronauts and the achievements of satellites have opened prospects scarcely conceivable a mere twenty years ago. The future depends on multidisciplinary competence, on close cooperation on an equal footing between the physicist, the mathematician, and the earth scientist.

PART 2

ARTIFICIAL SATELLITES AND REMOTE DETECTION

First, let us remember that we are dealing with earth sciences alone; meteorology, without being left out altogether, is of somewhat secondary importance here. It makes little sense to use consistent units where very large areas are concerned. Ground resolution must be compatible with our needs; but only the Nimbus satellites offer information per surface area unit of a little less than 100 km^2 directly below the spacecraft, a little more than 400 km^2 on the perimeters of the areas scanned. This is rather sparse data, but it is the best obtained up to 1974. On television, these surface units brought to some 10 to 20 km^2, which could satisfy us if it were not far more preferable to work with traditional aerial photographs or on plates brought back by the Gemini and Apollo astronauts.

Soviet, French, and British satellites are of limited use to us, although of enormous interest for basic research in meteorology and related sciences. Interest presently focussed on the earth sciences had just led to launching of satellites for research in that area. Let us hope that, at the international level, the cooperation already existing in meteorology will be extended to this domain, in which geographical sciences will occupy pride of place.

TYPES OF SATELLITES

Most satellites are either meteorological or military. The latter do not concern us at all, while the former offer us the opportunity to work on the data gathered. True, this opportunity is marginal, but it allows us to prepare for the future.

Table IX gives, as on 31 December, 1969, the statistical status of satellites launched throughout the world from October 1957 (Sputnik 1) to 25 December, 1969 (Intercosmos 2). This table includes 'non-identified' satellites (euphemism for spy satellites), but does not include those with no scientific payload*.

A number of Soviet satellites could certainly meet our needs despite very inadequate ground resolution; unfortunately, it seems that the information gathered by these meteorological satellites does not belong to the public domain, as is the case for Nimbus, ESSA, TIROS, and ATS**.

Only the following satellites offer suitable (and different kinds of) information: TIROS-ESSA, ATS, and, especially Nimbus (the ERTS satellites belong to the Nimbus type).

The above listing and Table IX have been stopped at the end of 1969. It would be tedious to keep going with this sort of 'comptability'. It is enough to say that, every year, space receives about 50 new satellites or probes, sometimes fewer, some other times many more. For instance, during the period 1 October 1971–30 September 1972, around 50 spacecraft have been launched; among them, 40 are publicly 'acknowledged', the rest belonging to the 'unidentified' category. Out of these 40 publicly recognized satellites, 18 belong to the U.S.A., 15 to the U.S.S.R., 2 to the United Kingdom, 2 to France (1 is shared with the U.S.S.R.), 2 to ESRO and, 1 to Japan.

1. 'TIROS-ESSA' Type

TIROS (*T*elevision *I*nfra*R*ed *O*bservations *S*atellite) and *ESSA* (Environmental Survey Satellite of the United States Meteorological Bureau, *E*nvironmental *S*cience *S*ervices *A*dministration) are well known satellites rotating on their own axes with the pet name – at the Meteorological Service – of Big Wheel. TIROS has its axis of rotation in the same position with respect to the plane of the ecliptic so that, during part of its orbit, it faces space, not Earth. On the other hand, the axes of the ESSA's are always directed towards the center of the Earth and, thus, always face our planet, which fills the whole field of view of the instruments.

* See Appendix A: slightly more detailed but still fragmentary table on the 'great firsts in space.'
** See *TRW Space Log* **8**, No. 4, Winter 1968–1969.

TABLE IX

Number of satellites launched up to the end of 1969

Country or agency	In terrestrial orbits		Lunar missions		Solar orbits		Planet touch-downs	Total	
	Launched	Still orbiting	Total	Still orbiting	Total	Still orbiting		Launched	Still orbiting
Australia	1							1	1
Canada	3	3							3
ESRO	4	3						4	3
France	5	5						5	5
Germany	1	1						1	1
Intelstat	9	9						9	9
Italy	2							2	
U. K.	4	4						4	4
United States	580	278	22	1	14	14		616	293
Soviet Union	380	77	15	4	8	8	4	407	89
Total	989	380	37	5	22	22	4	1052	407

N.B. 'Still orbiting' does not necessarily mean still working.

Ten TIROS satellites were launched; the first in 1960, and the tenth in 1965. With the exception of the last two, whose orbits were sub-polar, they were inclined 48° or 58° to the equatorial plane. All except four operate in the visible (TV) range; the exceptions are TIROS 2, 3, 4, and 7 which also work in infrared, but with very poor ground resolution. Though extremely useful for work in meteorological sciences, these satellites are of no interest to us.

The ESSA satellites, nine in all, (the first launched on 3 February, 1966, the ninth on 1 April, 1969), are all sub-polar. All carry TV cameras and four also work in the infrared, but with unacceptable resolution for earth sciences (ESSA 3, 5, 7, and 9).

All of the Tiros became inoperative the same year or the year after they were launched. The same applies to the ESSA's, except ESSA 2, 6, 8, and 9 which were still transmitting information in 1970.

2. ATS Type

The *ATS* (*A*pplications *T*echnology *S*atellites) are called 'stationary', i.e., they perform one revolution about the Earth in the same time that it takes the Earth to rotate once about its axis. They practically follow the equatorial plane, being inclined 0.2° (ATS 1) and 0.4° (ATS 3) to the equator. They can be used as relay satellites: for instance, ATS 3 was used to relay the television broadcast of the Mexico Olympic Games (video channel only, the audio used another satellite).

ATS 1, launched 6 December, 1966, transmits in black and white; ATS 3, launched 5 November, 1967* transmits three B&W images, transformed into a color picture, on the ground. Both transmit a picture of the part of the Earth in view every 25 min. With the records received from ATS 1 and 3 it is possible to make up excellent cinema photographic montages to show cloud movements.

ATS 1, is generally situated over the Pacific, while ATS 3 is usually over the Atlantic. The word 'generally' means that it is possible to place them above different points of the Earth as needed.

Earth scientists have little interest in the images received from the ATS. However, the ATS of the future will carry infrared radiometers whose space resolution of 0.1 milliradian will permit ground resolution of 3 to 4 km (altitude 30000 to 40000 km).

3. 'Nimbus' Type

The Nimbus satellites (named after the Latin word for 'cloud'), like the manned satellites, are the only type of definite interest to earth scientists. As distinct from all other satellites, the Nimbus do not rotate around their own axis. Throughout their successive orbits, they maintain an attitude that could be called motionless if their speed were not 24000 km/h. The satellite is 'Earth-oriented' and is three-axis stabilized.

* ATS 2 was launched in April, 1967, ATS 4 on 10 April, 1968. Both were shortlived due to malfunctions.

Nimbus 1 (A), launched on 28 August, 1964, had the short lifespan of three weeks. It is the only satellite whose elliptical orbit had an apogee of 930 km, and a perigee of 422 km, due to a malfunction in the launching rocket (third stage). It carried out three different experiments.

Nimbus 2 (C), launched on 15 May, 1966, lived, on television, until 1968. It ceased infrared transmission from November 1966 on, due to a malfunction of the onboard taperecorder. It carried out four different experiments. Its orbit oscillated between 1095 and 1178 km.

The launching of Nimbus B failed on 18 May, 1968. Launched successfully on 14 April, 1969, under the name of Nimbus 3 (B), this satellite performed nine different experiments. In 1970, it was still transmitting 'directly' but, once again, serious defects in the onboard taperecorders prevented quantitative analysis of the information received from November 1969 on. Its orbit is almost perfectly circular at 1111.2 km altitude.

Nimbus 4 (D), launched 8 April, 1970, travels at 1111–1112 km; it performs eight experiments.

Nimbus 5 (E), launched in December 1972, resembles the preceding Nimbus except for a few scientific instruments, among them the 'SCMR', dual radiometer operating on both sides of the local minimum, and having very fine resolution (0.6 mrad). Unfortunately, after transmitting a few splendid images, of Florida for instance, this radiometer went out of order (see Plate D, No. 3). Also, for the first time a passive microwave radiometer is operating in the 1.55 cm wavelength; unfortunately, its resolution is not fine enough for the earth sciences.

Nimbus F (Nimbus 6?), a 'carbon copy' of Nimbus 5, is scheduled for 1975–1976.

ERTS 1 (A), launched the 23rd of July 1972 is the first Nimbus-type satellite that is not meteorological, but exclusively devoted to the earth sciences. ERTS B (ERTS 2?), perhaps a carbon copy of ERTS 1, perhaps carrying the same Multi-spectral Spectrometer (MSS) with the fifth channel, thermal infrared, is scheduled for 1974–1975. Perhaps sooner?

The 100° angle of inclination to the equator, the period of 107 to 108 min, and the orbital altitude of these Nimbus satellites permit them to vary from orbit to orbit by 26° longitude towards the west. They always pass over given points on the Earth at the same local time, overflying them at least twice every twenty-four hours, once in the daytime and once at night.

The areas covered by the Nimbus' overlap slightly from one orbit to the next, the overlap becoming greater as higher altitudes are approached. In twenty-four hours, somewhat more than thirteen orbits are completed. The ERTS scanning is restricted to 100 N.M. (180 km).

The hour window (time over the equator) for the Nimbus is about 12:00 noon and 12:00 midnight, a time chosen because of the spectral quality of midday solar radiation. ERTS 1 crosses the Equator at 9:30 and 21:30.

Unfortunately, the difference between real solar time and equator flyover time becomes ever greater as higher altitudes are reached. This very important phenomenon will be discussed in the next chapter.

CHAPTER I

THE PROBLEM OF REAL TIME

The characteristics of subpolar orbits are such that time differences are much greater than most people think. Two factors must be borne in mind, one relative to the zone swept from west to east by the satellite, the other resulting from the longitudinal displacement of the satellite.

1. Opposition Between Eastern and Western Margins (does not apply to ERTS or to one Nimbus 5 radiometer)

The zone traversed and usable extends for about 30° longitudinally, corresponding to a time difference of two hours. Other things being equal, it should be understood that the thermal contrasts are quite large and, in the daytime, the energy reflected cannot be compared between the two extemes of the zone covered, since the angle of incidence of the Sun's rays at the two extremes differs too much.

1.1. Daytime

(a) *Thermal energy*. Consider two points spaced 30° longitude apart. When it is midday at the eastern point, it is only 10:00 am at the western point; in other words, the solar energy accumulated is geater at midday than it is two hours earlier. Thus, it is difficult to compare earthly phenomena at areas so far apart.

(b) *Reflected energy*. This phenomenon is perhaps even more important. To reach western regions, the optical path traversed by radiation through the atmosphere is so long that, under the right circumstances, the spectral quality there would be inferior to that normally expected. Also, as scattering affects the short wavelengths, it could even reach the near infrared if the particles in suspension were large enough.

1.2. Nighttime

Practically the same situation could occur since radiation picked up by the instruments on board the satellites is a function of that part of the energy accumulated during the daytime that is reemitted during the night. This nocturnal radiation begins as soon as the Sun sets, i.e., two hours earlier in the east than in the west.

1.3. Comparison for ascending and descending orbits

Figure 9 was derived from a large number of Nimbus 2 orbits. To make things easier, I have assumed that the satellite crosses the equator at midday and midnight.

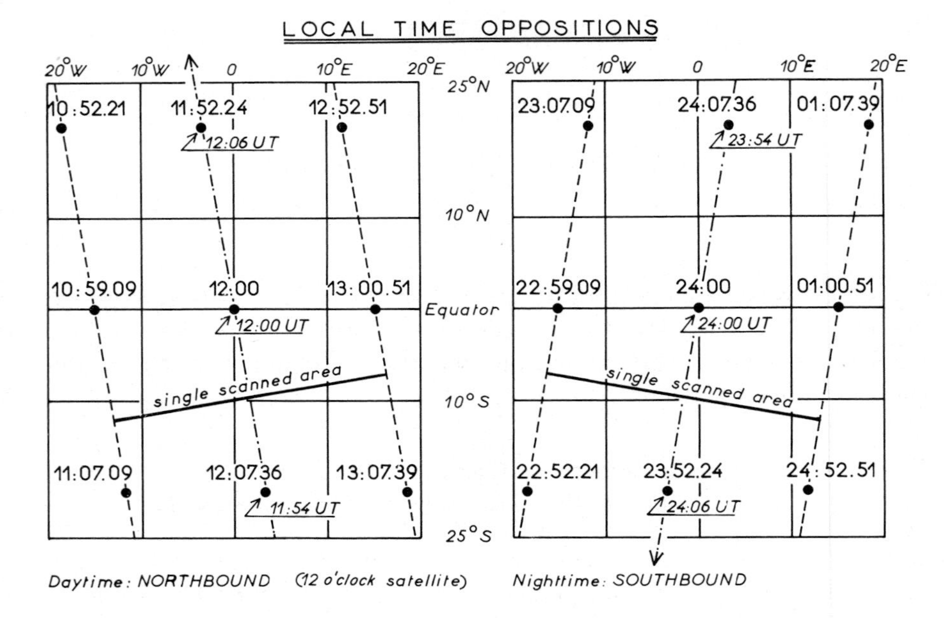

Fig. 9. Time differences between the opposite margins of areas scanned during one orbit (does not apply to ERTS and Nimbus 5 'SCMR') – The dashed lines delimit the eastern and western edges. The center line corresponds to the orbital axis as projected on the ground. On the left: northbound flight during the day, the equator passed at midday at 0° longitude (ascending mode). On the right: southbound (nightime), the equator being crossed at midnight at the same longitude (here arbitrarily defined as 0°). The boldface line perpendicular to the line of flight corresponds approximately to a single scanned area. UT = Universal Time, i.e. GMT.

(a) *Nighttime.* During its nocturnal trajectory, the satellite is going from north to south*, or more precisely, from 10° E north to 10° W south. I have noted the crossing times at 20° N, 0°, and 20° S. A curious but logical phenomenon will be observed: the satellite seems to be 'going backwards' as compared to local time, but the real situation must be read in GMT (Greenwich Mean Time) or UT (Universal Time).

Each area scanned by the instruments on board the satellite is obviously perpendicular to the axis of orbital motion, the two margins being at different latitudes. When a map is made of the data received from the satellites, the time difference is increased by about one minute between two points situated at the same latitude on either side of the map (from E to W), the point in west having been 'seen' a little later than its corresponding point in the east. This fact is recorded for general information only, as it has no practical importance.

* All Nimbus satellites operate in the same way. Decending node: longitude at the equator through which the satellite passes at night. Ascending node: longitude at which the equator is crossed during the day.

For practical reasons (spectral quality of solar radiation) exactly the reverse path was chosen for the ERTS satellites; i.e., north-south direction during the day so that the northern hemisphere (continental hemisphere) could be observed at convenient hours, the hour window occurring at 9:30 .

(b) *Daytime*. The situation is exactly the same in the daytime as at night. However, as we will see, if the time differences are the same, interpretation is a delicate matter.

1.4. Practical consequences

(a) *Nighttime*. Experience, including ground surveys which I made for multiple consecutive 24-h series, has shown that this difference can be ignored during the night, at least at the middle latitudes, for the so-called twelve o'clock satellites (12–24 h). At 23:00 or 1:00, nocturnal cooling has practically reached the same level, and the radiation temperatures at both margins of the area scanned can be validly compared. In particular, some minor adjustments have to be made. I need only cite the cases of water and sand.

Water gives back accumulated energy by radiation only very slowly, which is rather fortunate for the principles of interpretation. On the other hand, sand cools very fast, reaching relatively low temperatures in a few hours – well before midnight or 11:00 pm.

(b) *Daytime*. At midday the sand is very hot while the water is cold, but the water is relatively less cold in the west than in the east, as compared to the state reached by sand. Do not forget that by 10:00, objects are already warming up. Two hours later the differences are even more marked.

This example (sand-water contrast) is by no means an isolated example, and should encourage us to exercise caution when examining daily values. I have just said that the water-sand example was not extreme – far from it. Fieldwork analysing the diurnal radiation temperatures of Nimbus 4 has found extreme differences in the behavior of rock material according to its color, structure, state of disintegration, etc. This work has proved that, between 10:00 am and midday, quartz, and schist side by side react thermally in very different ways (Pouquet, 1972, *Rev. Géom. Dyn.*).

It is essential to bear these details in mind to avoid flagrant errors of interpretation. Perhaps we should confine our investigations to regions whose longitudinal stretching does not exceed about 15°; i.e., to regions with acceptable time increments. This suggestion applies only to radiance temperatures obtained in the daytime, and loses importance for information gathered around midnight.

2. Time Increments as a Function of Latitude (does apply to ERTS)

This is the most insidious problem of all, since it has been completely neglected by most users. Instead of working with an arbitrary scheme as before, it would be more useful to base my explanation on concrete facts borrowed from Nimbus 3, and on an isolated example from an ESSA satellite.

On 1 May, 1969, pass No. 234*, northbound in the daytime, crossed the point

* This is in fact orbit No. 234 north of the equator and No. 233 for the part of the orbit before the equator crossing. It has been (arbitrarily) decided to number the orbits according to their ascending pass above the equator. The word 'pass' is more correct, based on successive orbits since the moment of launching. In fact, 'orbit', 'pass', and 'swath' are used with the same meaning at both NASA and ESSA.

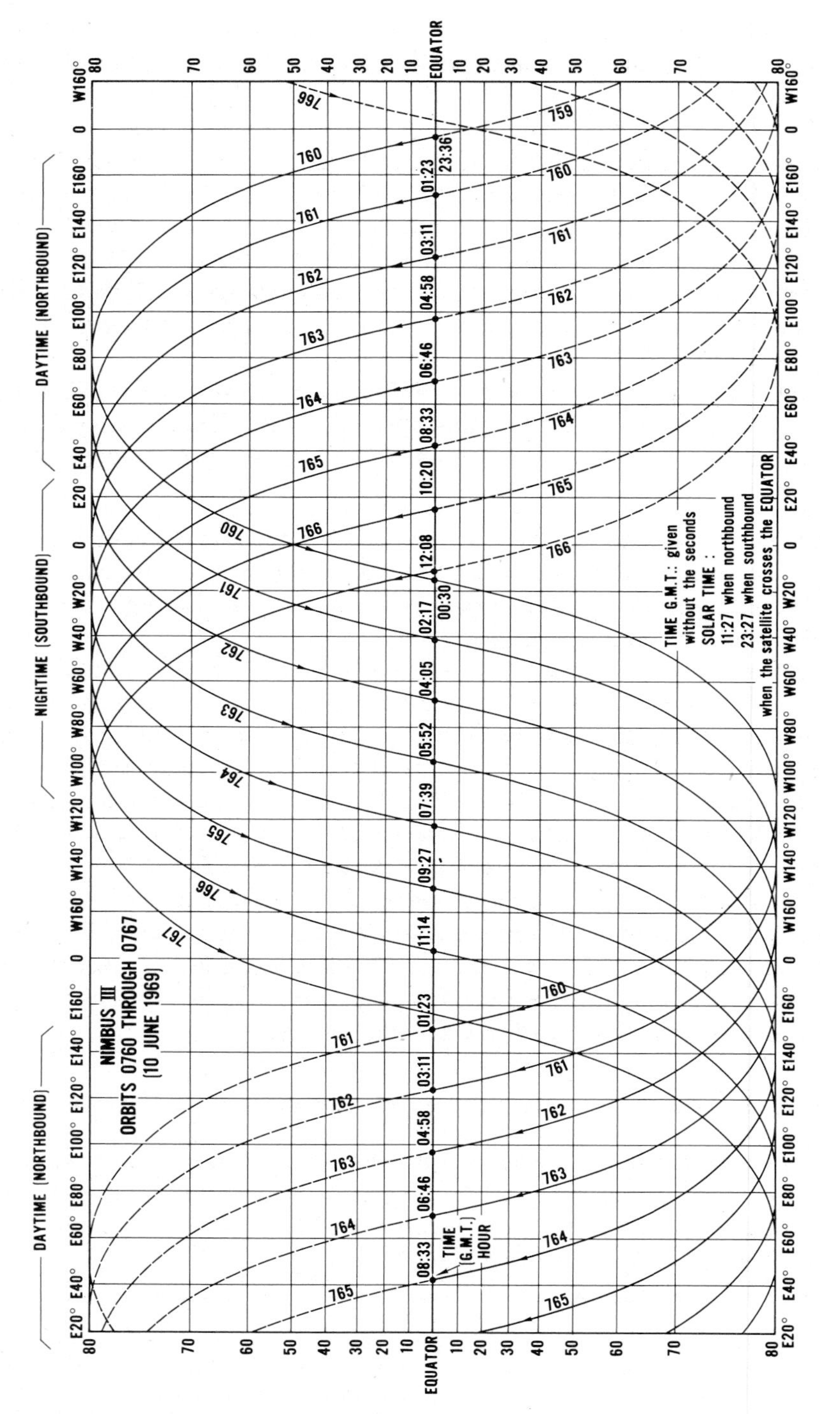

Fig. 10. Orbits Nos. 760–767, 10 June 1969 (Nimbus 3). – The GMT time is given at the point of each pass over the equator, the local solar time being 11:27 (11:52 for Nimbus 4). The arrows show the southbound direction during the night (center of figure); northbound in the day is on the left and on the right.

2° N and 99° W at 11:27 local time; at 10:35 local time, but in fact 16 min later, the satellite passed through 54°.2 N, 116°.7 W; at 5:58 local time, or 18:29 GMT, it was at 80° N, 172°4 W.

Here is another pass, no. 1579, on 10 August, 1969, still during the day:

at 11:23 local time (1:41 GMT), position 3.6° N, 145.6° E
at 10:25 local time (1:57 GMT), position 56.3° N, 127.2° E
at 4:05 local time (2:07 GMT), position 80.1° N, 44.6° E

At nighttime, the extremes of local time from high to low latitudes, are:

pass 234: 1:36 (18:34 GMT) and 23:28 (18:56 GMT)
pass 1579: 3:53 (2:08 GMT) and 22:28 (2:33 GMT)

If one recalls that Nimbus 3 crossed the equator at exactly 11:27 and 23:27, local times, it is easy to calculate latitude and longitude, which is not given here for this nocturnal example.

Figure 10 uses all the parameters relative to orbits 759 through 766. This map is quite precise, and can aid better understanding of the importance of the time problem resulting from the convergence of longitude with increasing distance from the equator. For example, for Nimbus 2 through 5.

(a) from the equator to 30° latitude, the total difference is 8'.37", i.e., 2'.8" per 10 degrees of latitude;

(b) from 30° to 60° latitude, total difference is 14'.16", or 4'.72" per 10° latitude;

(c) from 60° to 70° latitude, the difference is 12 minutes per 10° of latitude.

We will conclude with a final example. I deliberately took an operational satellite, ESSA 8, whose period is 117 min (instead of 108) and whose inclination is 98° (instead of 100°). To simplify the matter, I assumed that the satellite, instead of being a 'nine-o'clock satellite' (crossing the equator at 9:00 and 21:00) is a twelve-o'clock satellite, like the Nimbus, which in no way changes the problem. Table X uses the original parameters.

TABLE X

Longitude and time differences of ESSA 8

Latitude	Longitude	GMT time (UT)	Local time (solar time)	Time differences W/respect to UT (GMT)	Time differences W/respect to equatorial time
0°	0°	12h	12h	0	0
10° N	2°30' W	12h 3'15"	11h 35'15"	−10'	− 6'45"
20° N	6° W	12h 6'30"	11h 42'30"	−24'	−17'30"
30° N	9° W	12h 9'45"	11h 33'45"	−36'	−26'15"
40° N	13° W	12h 13'	11h 21'	−52'	−39'
50° N	18° W	12h 16'15"	11h 4'15"	−1h 12'	−55'45"
60° N	25° W	12h 19'30"	10h 39'30"	−1h 40'	−1h 20'30"
70° N	41° W	12h 22'45"	9h 38'45"	−2h 44'	−2h 21'15"

Having reached the latitude of 70° N, ESSA 8 'brushes against' 82° N and begins its descending nocturnal trajectory. It reaches the same latitude 70° N at 2:21′15″ solar time, passes 40° N at 0:39′, 20° N at 0:17′30″, the equator at midnight, 10° S at 23:53′15″, etc.

In the case of the Nimbus series, it is possible to draw Figure 11 for Nimbus 3 (equatorial nodes at 11:27 and 23:27). However, the time differences have been

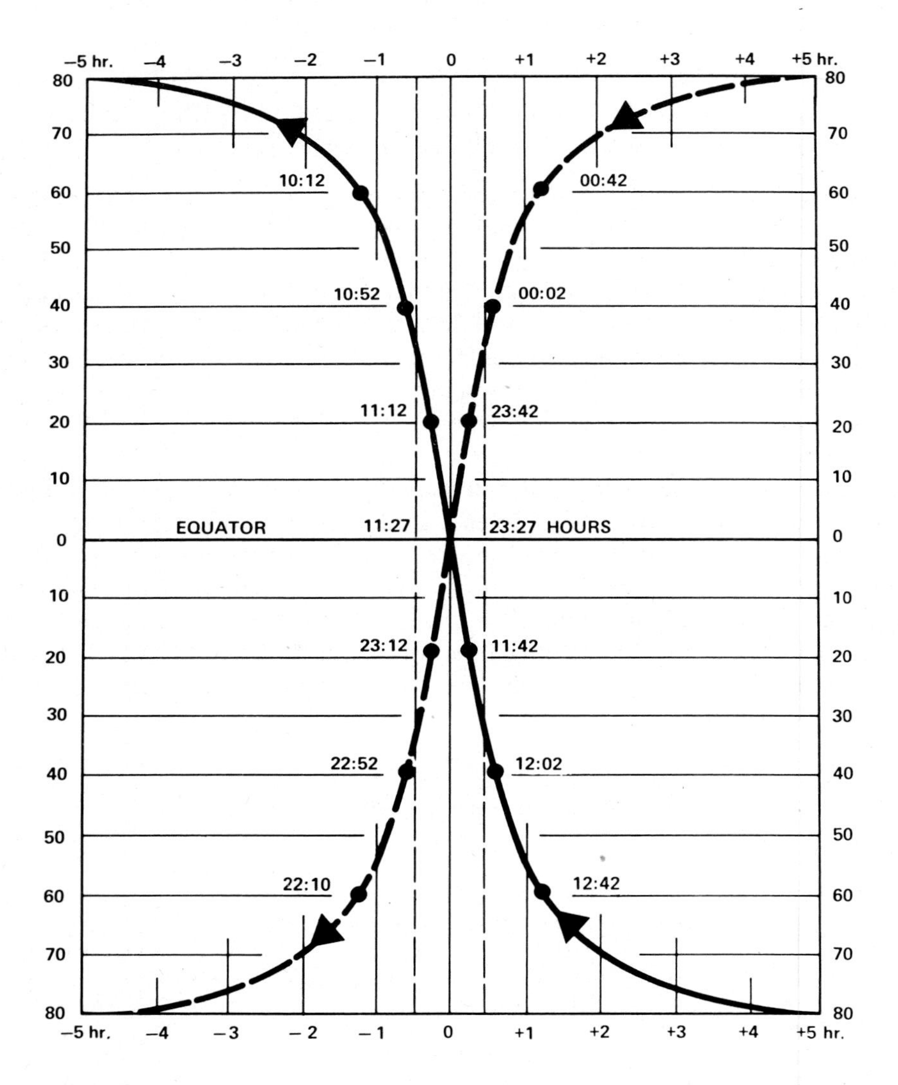

Fig. 11. Solar time directly below the satellite as a function of latitude (applies to all the Sun synchronous satellites, Nimbus, ERTS ...) – *Solid lines:* diurnal trajectory; *dashed lines:* nocturnal trajectory. This graph was drawn for Nimbus 3. For Nimbus 4, add 25 min to all hour figures. Taking 0 as the time when the satellite crosses the equator, the negative and positive numbers at the top and bottom of the figure give the time differences.

designated with respect to a time 0, which is the time the satellite crosses the equator. The period, altitude, and inclination of Nimbus 3 allow for a very slight lessening of time differences observed for ESSA 8. The orbital characteristics of the latter Nimbus and the ERTS can be easily derived from the graph – Figure 11 (however, note that the ERTS are southbound during the day).

3. Importance of the Real Time Problem

The size of the time difference obviously depends on whether the time for objects situated at different longitudes is earlier or later, which consequently amplifies the differences with respect to the inclination of solar rays striking these same objects. Taking the example of orbit 234 of Nimbus 3, and that of orbit 1579, the solar angle, directly below the satellite, was:

1 May, 1969:	74.8° by 2.0° north latitude
	47.8° by 54.2° north latitude
	15.0° by 80.0° north latitude
10 August, 1969:	74.1° by 3.6° north latitude
	45.3° by 56.3° north latitude
	12.0° by 80.1° north latitude

This leads us to consider four aspects relating to satellites.

3.1. Records of the visible (tv cameras)

The impact in the visible could appear negligible, since the ground resolution of the TV cameras does not allow the shadows cast to be distinguished individually. The lenses on board give an average of the reflected radiations, however, and it is evident that these shadows tend to distort the spectral quality of the reflected light. However, I must once again repeat that televized records are of secondary interest in earth sciences; aerial photographs and photographs brought back by the astronauts are so much better. To be quite frank, I have never examined these televised records aside from mere curiosity, or to check for the absence (or presence) of low cloud ceilings.

3.2. Reflected near infrared (nimbus 3, 5 and erts)

We must examine the circumpolar regions and the lower latitudes separately.

(a) *Middle and low latitudes*. At first sight, the possible error is rather large, since objects receiving the Sun's rays at an angle of 50° can hardly be compared with those receiving these rays at an angle of 25°. In fact, the possible error is partially eliminated by conversion of the recorded energy into the infrared albedo allowing for, albeit imperfectly, the latitude-longitude position, the time, the day, and the month of the shots. Unfortunately, corrections made for real time are somewhat rudimentary. By taking the above into consideration, the interpreter will avoid making gross errors.

(b) *Circumpolar regions*. In these regions, the exact time is of trifling importance. On the other hand, the extreme obliquity of the solar rays causes considerable lengthening of the shadows, which affect the reflected radiation. Whereever these polar landscapes show high relief we see low-reflectance fringe areas which might lead one to believe in nonexistent phenomena, for instance the presence of free water on the Antarctic glaciers.

3.3. Thermal infrared

We already know the data of the problem. I would like to add just one point to a very simple but often disregarded aspect of the problem.

As the rates of diurnal heating and nocturnal cooling differ from one object to another, it is inevitable that the curves showing these rates will intersect. These intersections complicate the task of the interpreter: if for no reason other than this, we can understand why the choice of a time window for the satellites is a delicate subject.

3.4. Optical time for launching satellites

We touched on this subject when we considered the best time of the day and season of the year for getting the best photographs. In fact, the same conclusions apply to satellites in both the near and middle infrared and visible ranges.

We need to have clear differentiations between the objects under study. We need to know that this or that formation (plant, rock) usually has a radiance temperature higher or lower than that of its neighbor. Above all, we must do everything possible to avoid confusions that can result from the intersections mentioned above.

My numerous field missions in Canada (low thermal differences in the fall) and in the southwestern deserts of the U.S. (high thermal differences) consistently showed: (a) untimely intersections between 8 and 10 then 19–21:00; (b) sharp differences between 11 and 14:00; (c) minimal differences around 5–6:00; (d) it sometimes, though rarely, happens that intersections are observed between midnight and 2:00 and around 12–14:00.

Under these conditions, the twelve-o'clock satellites seem to be the best. On the other hand, if we remember the longitudinal convergence towards 50° of latitude and beyond, we are plunging into the heart of unfavourable time schedules.

In conclusion, the ideal would be to *launch at least three satellites* in the respective hour windows: 8–10, 11–12, and 14–15:00. For a given region, it would be enough to choose the one corresponding to the time needs, simply by consulting Figure 11. Moreover, these three satellites, overflying a given region two to three hours apart, would enable us to acquire invaluable knowledge of the tendencies shown by the various objects studied, their rates of cooling and of heating.

Three satellites would give six readings in twenty-four hours. In the very near future we could combine the information from these three satellites with that from the orbiting Nimbus, ERTS, and ATS when these are equipped with ultra-high resolution infrared radiometers. This would give us 8 to 12 readings per twenty-four hours, and would enable us to perform essential statistical processing. The chief dynamic and situational parameters (geometric mean, median, interquartile ranges, slope of quartiles, asymmetry, standard deviation, etc.) all can be used to identify terrestrial formations in an absolute manner. One need hardly add that these multiple 'identity maps' could be stored in the memory of a computer which would thereafter provide solutions to problems addressed to it.

CHAPTER II

INSTRUMENTATION ONBOARD THE SATELLITES

1. Types of Instruments According to Spectral Band

The scientific payload to be placed on board satellites (or aircraft) depends on two types of considerations: (1) the target in view and (2) the technical means available.

1.1. Targets

Table XI was prepared from concrete data but is far from being complete, or even accurate, since too much uncertainty exists throughout most areas. However, imperfect as it is, some practical inferences may be drawn.

Leaving aside cases full of query marks, thermal infrared comes out clearly at the top, followed by infrared film (unusable on satellites at present). In the first and fourth places come passive microwaves* and reflected infrared, which are out of range of photographic emulsions. Next to the last is radar, followed by the visible range (photography and television). This order would probably be somewhat rearranged if Table XI were to include multispectral photography as it exists on earth science satellites (ERTS).

1.2. Method of operation

Table XI actually proposes a fairly loose choice of logical methods of operation. Taking the spectral bands usable for specific purposes, we may establish the following list:

(a) *Reflectivity studies* (clouds, earth, sea) from about 0.4 to 2.5 μm, using the complete range of possible instruments.

(b) *Thermal blackbody values* (clouds, sea, earth), atmospheric windows 3.5–4.2 μm and 8–14 μm. Complete range of instruments excluding TV and other cameras.

(c) *Brightness temperatures* (clouds, sea and glaciers, earth), radiometers predominate (passive microwave).

(d) *Meteorological research* (atmospheric water vapor, ozone, ultraviolet). Use of radiometers and spectrometers.

* When we achieve linear ground resolution equal to that of the infrared radiometers the passive hyperfrequencies will occupy the first rank.

TABLE XI

Possibilities for remote detection of terrestrial phenomena from a satellite, an aircraft, or on the ground

Terrestrial phenomena	Visible	Infrared			Microwave		Observations
	TV, Film	Film	Reflected not phot.	Thermal D = Day N = Night	Active (radar)	Passive	
Forms of terrain	*	*	*	D *N	*	*	Continental and ocean masses; shapes, etc.
Soil erosion	*	*	*	D *N	*?	*	Phot., TV: gully-erosions IR: truncated profiles
Soil humidity	/	*	*	D *N	*	*	
Old watercourses	/	*?	*?	*N	?	*?	
Altered rocks, ancient aeolean sands	/	*	*	D *N	/	*?	
Very fine material (clay, sand ...)	/	/	/	/	*	/	Side-looking, wavelength at least twice particle size
Deep, narrow gorges	*	*	/	*N	*	*?	Radar, side-looking: photo, shadows cast, IR, therm. contrasts
Tectonic features, faults, fractures	?	*	*	*N D?	*	*	Visible: when topographic feature
Geological forms (structure, domes, folds)	?	*	*	D *N?	*	*	Visible: with structural topographic forms
Currents (oceans, lakes)	/	(rare) *?	(rare) *?	*N	/	*	
Water pollution	/	*	*	D *N	/	*	
Plant life (stage of growth)	/	*	*	D *N	/	*	Harvest prospects
Mineral resources (surface)	/	*?	*?	D *N	?	*	
Industrial activities	*	*	*?	D *N	?	*	

Table XII leads to a new approach.

This table was prepared from the viewpoint of direct image production. In practice, radiation detected by the instruments, i.e., on a satellite or aircraft, is recorded on magnetic tape which is played back on command and recorded at the receiving station (some spectral bands are broadcast 'live' from the carrying platform (satellite)). These ground recordings are used for production of pictures and numerical data.

When all is said and done, TV cameras and, especially radiometers are the most useful instruments to the earth sciences. On satellites, scanning from horizon to horizon or over shorter distances is done by a mirror inclined at 45° whose speed of rotation is governed by the orbital velocity of the space vessel and the resolution of the instrument in question*.

2. Ground Resolution

2.1. General

Let us recall that the linear ground resolution is derived from the spatial resolution, in radians, and the altitude. In the thermal infrared the spatial resolution depends upon the angle of view. For an angle of 1°, the spatial resolution is 17.45 mrad; for 0.3°,

TABLE XII

Detectors that produce images, as a function of frequency

Spectral regions		Frequencies (Hz)	Detectors
Microwaves	10 –100 cm 1. – 10 cm 0.1– 1 cm	3×10^9 – 3×10^8 3×10^{10}–3×10^9 3×10^{11}–3×10^{10}	Scanning antennas and radio receivers
Far and middle infrared	1 mm–3 μm	3×10^{11}–10^{14}	Detectors with sweep system. Various types of image formation on CRT (cathodic tubes)
Near infrared	0.7–3 μm	10^{14} to 3.85×10^{14}	Photo. emulsions up to 1 μm and IR detectors with sweep systems. Excellent results with CRT.
Visible	400–700 mμ	7.5×10^{14} to 3.85×10^{14}	Films. Detectors with sweep system (multispectral), television.
Near UV Middle UV	315–400 mμ 300–315 mμ	0.952×10^{15} to 7.5×10^{14} 10^{15}–0.952×10^{15}	Films (quartz lenses). Detectors with sweep system. Image converter tubes.

Atmospheric opacity below 0.3 μm.

* It is not my intention to describe the phenomenon of detection or the detectors used; here, I recommend the reader to consult specialized texts. I will note only that the detector transforms the energy detected into electrical signals. Progress has been so rapid that detectors used in 1950 seemed obsolete in 1960 and antiquated in 1970.

4.235 mrad.... For the passive microwaves, the beamwidth, in radians, is the ratio wavelength/antenna diameter, multiplied by the constant 1.2.

At an orbital altitude of 1200 km and an angle of view subtending 7 mrad (0.4°), ground resolution is $1200 \times 0.007 = 8.4$ km. The [orbital] velocity of the satellite being some 403 km per minute, the speed of rotation of the mirror is thus $403/8.4 = 48$ rpm, a reasonable speed permitting excellent recordings. An eclipse shutter (chopper) analogous to that placed in front of the lens of a movie projector, cuts up the band scanned from left to right into individual segments (samples) whose width, from directly below the satellite, is equal to that of the resolution. Each of these samples is 'read' for about one-thousandth of a second.

The revolving mirror picks up the radiations emitted or reflected, and directs them, via a set of concave mirrors, to the detector supplied with the appropriate filter, thus permitting the desired wavelength to pass through. The signal is then amplified electronically and produces an image on a cathode ray tube; this image is automatically recorded on film outside the satellite. On space vessels, the amplified signal is recorded on magnetic tape then retransmitted to Earth, as we already know.

At this point I feel it necessary to note a very commonplace phenomenon. The mirror sweeps from left to right, or from horizon to horizon. Only the points situated directly below the satellite could, strictly speaking, be aerially photographed. In fact, the rotation of the mirror permits multiple individual photographs to be recorded along a swept line, the photographs (samples) becoming more and more oblique. The photographic data, when finally reconstructed, are thus made up of many 'photographs' (samples), vertical in the center and oblique on the margins with varying degrees of obliqueness in between. We should also recall that the reflected near infrared, and even more, the thermal infrared, have very little relationship, except one of origin, with the reflected light picked up by the photographic emulsion. *"If these ideas are not understood, the single advantage of infrared reconnaissance cannot be obtained, and many interpreters can conclude that the infrared image is nothing more than an aerial photograph with poor resolution qualities".* (Avery, 1968, p. 145; italics mine.)

These poor resolution qualities become worse and worse as we leave the vertical. As a function of the angle of view (nadir angle):

At 0°, the resolution delimits a surface of 7.7×7.7 km (Nimbus 3)

–15°, the resolution delimits a surface of 8.0×8.4 km
–30°, the resolution delimits a surface of 9.1×11.3 km
–40°, the resolution delimits a surface of 10.7×16.3 km
–50°, the resolution delimits a surface of 14×31.8 km

The above values show that, beyond a nadir angle of 30°, differentiation between terrestrial features becomes more and more difficult. Computers should thus be programmed for a maximum of 30°: in fact, as we were looking mainly for rules of interpretation and not definitive results in a geological and geographical context, we programmed to a maximum nadir angle of 50°. Since 1972, the problem has no longer arisen with Nimbus 5 and F and with the ERTS satellites, since scanning is limited to a nadir angle very much less than 30°.

1. Earth has turned from observer into observed.

(See detailed captions for plates on the facing page.)

2. Hurricane Camille, August 1969.

2.2. Types of resolution

To simplify calculation, we will assume that the orbital altitude is 1000 km instead of 1100 to 1200 km. The values we get will thus be slightly larger than the real dimensions.

Resolution can be *very low, low, medium, high, very high, ultra-high*, etc.

(a) *Very low, low, and medium resolution* do not concern us as they are used for meteorology, and upper atmospheric and stratospheric studies. The cutoff point between medium and high resolution can be set arbitrarily at 20 mrad, which gives a ground resolution of 20 km. When the resolution is expressed in degrees, it will be low and very low resolution which, on the ground, is about 87 km for an angular aperture of 5° (0.08726... radians).

(b) *High resolution* currently defines the instant field of view as less than 0.01 radian. Usage has condensed *H*igh *R*esolution *I*nfrared *R*adiometer into the acronym *HRIR*, which, in turn, was abandoned in favor of *THIR* (*T*emperature, *H*umidity *I*nfrared *R*adiometer) for Nimbus 4 and 5. High resolution no longer appears, but this is what it is, at least in the 11.5 μ channel.

As long as we are deploring these more or less whimsical abbreviations (don't forget the origin of RADAR!) I will report that on Nimbus 5 and F one says, not HRIR also THIR but – why not? – SCMR. True, this instrument has very high resolution.

(c) *Very high resolution*. High resolution extends roughly from 10 to 1 mrad. From 1 to 0.1 mrad it is usual to say very high resolution (VHR). The linear units delimited on the ground go from 1 km to 0.1 km. The SCMR to which I have just alluded operates at 0.6 mrad, and thus, directly below the satellite, distinguishes areas of 600 × 600 m.

(d) *Ultra high resolution*. Below 0.1 mrad, one might conceive of instant fields of view of 0.01, or even 0.001 and 0.0001 mrad which, converted into linear values on the ground, works out at resolutions of 10 m, 1 m, and 1 cm.

Is this fantasy? Having personally been a witness to technical marvels I will be very careful about answering this question. Rather, I will quote a poster hung by one of my colleagues over his desk at Goddard: "If you want the impossible, wait a little. Miracles take longer." It is hardly necessary to add that this colleague was often called upon to help solve ever-thornier problems.

Plate A.

1. Earth has turned from observer into observed. Plate from Apollo 8 in lunar orbit. The Earth may be seen on the horizon.

2. Hurricane Camille, August 1969. This hurricane caused considerable material damage on the coasts of the Gulf of Mexico. We can clearly see the 'eye of the storm' and a ring surrounding the central part of this tropical perturbation. Plate from Nimbus 3, IDCS (television).

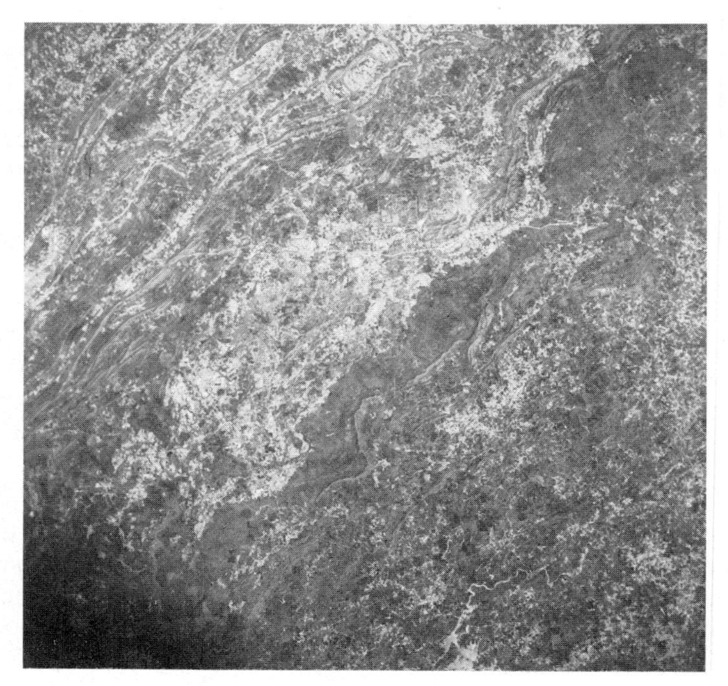

Photograph D (NASA code)

Multispectral photographs
Apollo 9.

Photograph C

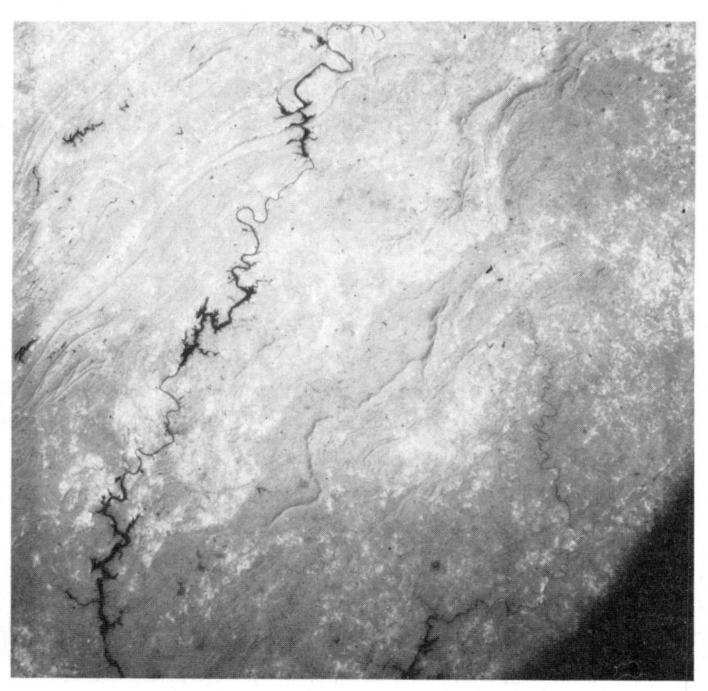

3. Equipment Onboard Satellites

The Nimbus satellites are experimental, not operational as are the ESSA. Consequently, each Nimbus is charged with checking working hypotheses and experimenting with new instruments. The results belong rather to the speculative sciences than to the applied sciences, but the long-term goal is to set up tested techniques for use on operational satellites. Table XIII gives an idea of how experiments are multiplying and improving.

For practical reasons, I have retained the NASA initials to facilitate the task of those who wish to write for information and documents. We already know the meaning of HRIR and MRIR (medium resolution); certain abbreviations are expanded in the table itself; but it will probably be useful to spell out AVCS, IDCS, APT, DRIR, and RTTS in full:

AVCS: Advanced Vidicon Camera System (3 TV cameras)
IDCS: Image Dissector Camera System (1 TV camera)
APT: Automatic Picture Transmission ('live' broadcast, 1 camera);
DRIR: Direct Readout Infrared Radiometer
RTTS: Real Time Transmission System.

For APT it is easy to install receiving stations cheaply (about eight to ten thousand dollars), that will receive transmissions directly (television, THIR) when the satellite is in line of sight.

In addition to the equipment listed in Table XIII, I must add that Nimbus 4 and 5 carry three other instruments (two spectrometers and one radiometer) for specialized meteorological experiments (atmospheric water vapor in the 1.2–2.4 μm and 3.2–6.4 μm band; ozone distribution, 0.25 to 0.34 μm band; thermal profile of the atmosphere and stratosphere, 13 to 15 μm bands, 6 different channels). To make the task of future researchers just that bit simpler, the instruments doing these various jobs are called, respectively FWS, BUV, and SCR. At the risk of repeating myself, I will say

---

Plate B.

Multispectral photographs, Apollo 9. Two photographs from the group of four, or C (black and white infrared, filter 89B) and D (Panatomic X, with filter 25A). The geological aspects, despite the plant covering, appear sharply on the two photographs, while hydrography shows up on plate C.

On Plate B, picture D, note U.S. Route 20 oriented west-east, connecting Birmingham (left edge, center) with Atlanta, Georgia (not on print).

From north to south, on Plate B, picture C, the Coosa river (Upper Alabama) with the dam in the center. The town of Gadsden is north of the dam and Anniston is to the east.

Center-right, bottom, northern part of Lake Martin on the Tallopoosa River.

Note the folds of the Appalachians crossed by rivers. The region is the central and northern part of Alabama.

Photographs taken 1st March 1969 at 16:21 GMT (10:37 local solar time) at an orbital altitude of 196 km.

AS9-3700 photographs. Photographs A (infrared, color, filter No. 15) and B are not reproduced in this book; A enables us to pick out a number of vegetation details, but B shows nothing new.

once again that earth scientists are only interested in using high and higher resolution instruments to detect features of the Earth's surface.

4. Signal Reception

To be brief, I will confine myself to giving some information on Nimbus 4, once more deploring the abuse of enigmatic initials.

4.1. Onboard the Satellite

Four independent systems are assigned to the various detection instruments.

TABLE XIII

Experiments performed on board 'Nimbus' meteorological satellites

Parameters and equipment	Nimbus A (Nimbus 1)	Nimbus C (Nimbus 2)	Nimbus B (Nimbus 3)	Nimbus D (Nimbus 4) etc.
Date of launching	24-8-64	15-5-66	14-4-69	8-4-70
Orbital altitude (km)	423–932 km	1095–1179	1,112	1,112
Period (minutes)	98.3	108	107	107
Inclination (to the plane of the equat.)	98.6°	100.3°	100°	100°
Television: Delayed	AVCS	AVCS	IDCS	IDCS
Live	APT	APT	APT	RTTS
HRIR: Live (DRIR)	no	no	yes	yes
Channel 3.5–4.2 μm	yes	yes	yes	no
Channel 10–12.5 μm	no	no	no	yes (THIR)
Resolution: mrad	7.9	8.7	7	7
ground (km)	7.5–3.5	9.7	7.8	7.8
MRIR: Channel H_2O	no	6.4–6.9 μm	6.5–7.3 μm	6.7 μm (THIR)
Window 11.5 m	no	10–11 μm	10–11 μm	see below
Channel CO_2	no	14–16 μm	14.5–15.5 μm	see below
Long wavelengths	no	5–30 μm	20–23 μm	see below
Albedo	no	0.2–4 μm	0.2–4 μm	see below
IRIS (infrared interferometer-spectrometer). Thermal profiles of atmosphere. O_3; water vapor; surface temperatures; secondary gases			5–20 μm	8–20 μm
SIRS (Infrared spectrometer); atmospheric temperatures at various levels			11–15 μm	11 μm 13–15 μm 19–36 μm
MUSE ('monitor,' surveillance of solar UV)			120 mμ 140 mμ 160 mμ 180 mμ 200 mμ 260 mμ	120 mμ 140 mμ 160 mμ 180 mμ 210 mμ 260 mμ

AVCS: three cameras. IDCS and APT: one camera.

(a) *HDRSS (High Data Rate Storage System)*. This large-capacity recording system is assigned to HRIR, MRIR, IDCS, and IRIS. The system itself is composed of two similar but independent assemblies, using five tracks on magnetic tape, one of them for the time code.

The recording speed of the magnetic tape is 3.5 cm/s. When queried, the satellite plays back the tape at 32 times this speed, or 112 cm/s. Ground transmission uses the S band, and the transmitter has a power of 4 W (frequency 1.7025 GHz). Each tape unit can record for 2h 14m, maximum.

(b) *PCM (TS) (Pulse Modulation Telemetry System)*. This system is assigned to MUSE and SIRS, and records then transmits the telemetric information. (In general the last two letters of the acronym are not used).

(c) *RTS (Real Time Transmission System)*. This is a direct transmitter assigned to TV and infrared (transmitter power 5 W, frequency 136.95 MHz).

(d) *IRLS (Interrogation, Recording and Location System)*. This apparatus which, as its name indicates, used when interrogating the satellite, is assigned to the various magnetic recorders and specifies geographic locations.

I think it unnecessary to stress that only the first system, HDRSS, is of interest to us.

For Nimbus 1 and 2 a single taperecorder was used (HRIR). Nimbus 3 inaugurated the method of doubling up recorders. Nimbus 1 ended its brief career with a malfunction of its solar panels. For Nimbus 2 through 5 the difficulties centered on taperecorder malfunctions. It is curious to note that the chief weak point in satellites occurs in the utilization of magnetic tape. The future – a close one we hope – will 'short-circuit' onboard recording and operate directly and continuously via relay satellites.

4.2. Ground Reception

Two receiving stations have been assigned to the Nimbus satellites, one in Alaska (near Fairbanks), the other in North Carolina (at Rosman). Fairbanks 'interrogates' the satellite for ten or eleven orbits and Rosman generally takes care of two orbits. Other stations record in real time for orbits 'missed' by the two main centers, and all receiving centers transmit directly to the Goddard Space Flight Center. Goddard, in turn, retransmits to the Weather Bureau (ESSA, Suitland, Maryland) and the SCR and THIR information, to the Clarendon Laboratories of Oxford University, Great Britain. Each station uses the information for its own purposes: for applied research (ESSA) or fundamental research needs (Goddard, Clarendon).

CHAPTER III

SATELLITES DESTINED FOR EARTH SCIENCES

> The Earth Resources Technology Satellite (ERTS) Program promises to be among the most significant undertakings in Space in the 1970's.
>
> The... Program is... a system devoted to developing and demonstrating the capability for more effective management of the Earth's resources (NASA and G.E., 1971, p. 1).

As we know, the earth sciences are the 'last-minute guests,' as satellites launched up to 1970 were essentially for meteorological purposes. The capability offered by high-resolution radiometers, however, made it possible for our discipline to participate in fundamental research.

In parallel, certain catchwords have become popular with the general public, such as air and water pollution, and environment, now associated with land use. This fact, along with the tremendous progress in meteorological research achieved through satellites, has finally squeezed out a trickle of funds to support new types of space vessels, and from these we expect momentous contributions to the fields of land development, environmental studies, and detection of air and water pollution. Thus, from being taggers-on, we now have 'full membership,' first with the Nimbus satellites from number 4 on, and then (in the same year of 1972) we are exclusive guests invited onto the Earth Resources Technology Satellites (ERTS). These, in fact, carry hardly anything but Earth science instruments; meteorological instruments are of secondary importance. Nimbus 5 and F continue to be meteorological satellites, but they each carry one instrument for which meteorologists have little use.

This introduction is a mere sketch of what is to follow: before we actually enlarge on the satellites reserved partially or exclusively for us, it is only fair that we pay homage to the work of the 'prehistorians' of our discipline, a category to which I am proud to belong. Since I am a NASA man by both training and preference, I am going to use the arcane symbols so dear to the hearts of my colleagues (but not all of them) by introducing each subsection in this chapter with these initials as headings.

1. NERO

NERO stands for *N*imbus *E*arth *R*esources *O*bservations, and is the title of Volumes I and II of a publication by the private company Allied Research Associates*.

To-day, the results obtained are more than satisfactory: to convince yourself of

* I should state that, as a 'scientific consultant' to this company intimately associated with NASA, I participated in the preparation of Volume II of NERO, which will be found in the bibliography under the names of my colleagues, R. R. Sabatini and J. E. Sissila.

this, just run down the non-exhaustive list in the bibliography under the heading *Practical Results*. You will note that, unfortunately, most of the publications based on satellite observations consist almost entirely of interpretations of TV photographs and facsimile records for infrared.

I fear that most of my colleagues have boggled at the mountain of work involved in reducing numerical data. I have to admit that it is fair from easy to get hold of these records, since NASA cannot keep up with the demand. As it happens, I have observed three characteristic attitudes on the part of earth science experts: (1) those who have no desire to go back to school and state categorically that remote detection, especially by satellite, is of no use; (2) the unbridled enthusiasts who state just as categorically that remote detection, especially by satellite, is everything and nothing else exists; and finally, (3) a tiny minority (to which I belong) who believe it necessary to go back to their studies and add another string to their bow without, on the other hand, neglecting the methods brought to a fine art and bequeathed to us by our predecessors.

In fact, we should admit that in-depth studies are very few and far between; only a handful of us have undertaken this kind on work on digitized maps. *Have we come up with positive results*? is a question often asked by geographers, and sometimes by geologists.

We have to come to some agreement on the meaning of the word. If by 'results' we mean something comparable to the findings obtained with classical methods by these geographers or geologists, then the answer is almost 'no.' But this is not the point.

Our sole ambition was to seek out the best working methods (what scale? what contour lines?...), to set up the parameters for valid interpretation, and to act as technicians so that when the satellites were orbiting, the road ahead would be clear, and the information received from SCMR, ESMR, RBV, and MSS would be more easily understood and deciphered. Considered from this viewpoint, our results, although incomplete, are promising. I believe that we have worked out a doctrine: i.e. logical bases for interpretation. To do this, we have had and always will have to overcome innumerable obstacles, among which I will cite: ready-made theories (generally wrong), ideas accepted without thorough checking, and, last but not least, the problem of scale (we will return to this problem at greater length).

2. SCMR and ESMR

I think we are by now quite familiar with HRIR and THIR. For Nimbus 5 and Nimbus F, the satellites with which we will be dealing, we should say THIR-SCMR and, less frequently, ESMR.

2.1. Meteorololical payload of nimbus 5 and nimbus f

The meteorological equipment is made up of three instruments, namely:

(a) *Nimbus E Microwave Spectrometer* (NEMS). Five channels are used, two for studying atmospheric water vapor and three for molecular oxygen.

(b) *Selective Chopper Radiometer* (SCR). Seven channels in the CO_2 band make up a vertical thermal profile with seven altimetric levels at 3, 9, 13, 17, 25, 30, and 41 km.

(c) *Integrated Profile Radiometer* (ITPR). This radiometer has seven channels; the two firsts correspond to the usual atmospheric windows.

2.2. Apparatus shared by meteorology and earth sciences

This apparatus is mentioned in the title of this section: ESMR (Electrically Scanning Microwave Radiometer), i.e., a passive radiometer operating in microwave bands with scanning provided by electrical sensitization of the receiving antenna. We hope for a good deal from this instrument, although its ground resolution is far from being satisfactory. This radiometer operates in the 1.55 cm band (19.35 GHz) and determines the amount of atmospheric humidity that is condensed into clouds. It also enables us to study the characteristics of the two polar icecaps, for which ground resolution of 45 to 50 km is usually sufficient.

2.3. 'Earth science' instrument: SCMR

The full name of the instrument is *H*igh *R*esolution *S*urface *C*omposition *M*apping *R*adiometer (HRSCMR). This somewhat cumbersome title simply means that the instrument has a very high resolution useful in mapping the Earth's surface*.

Two detectors receive, one in the 8.4–9.4 μm band, the other in the 10.2–12.4 μm band, avoiding the O_3 absorption region. The instant field of view is 0.6 mrad giving ground resolution of 600 to 630 m. The area scanned is limited to 400 km on each side of the field directly below the satellite: it thus takes three to four days to cover the area scanned by the old Nimbus (HRIR) in a single orbit.

As I see it, ground resolution is generally sufficient. Maps on the scale of 1:100000 can be obtained, allowing us to tackle our problems with unprecedented ease. However, the most important innovation is not that of resolution, but involves the placement of the two twin channels on either side of the dip (*Reststrahlen*) which has been discussed previously. In practice, the two detectors supply two radiance temperatures at the same instant for the same object. The thermal difference between these two channels, resulting from a difference in emissivity, enables alkaline rocks to be distinguished from acid rocks. If no difference is recorded (plant cover, snow, oceans) we get, as in the past, a radiance temperature distribution map.

This is considerable progress over the situation of a few years ago. Provided we have a large number of investigators familiar with these techniques, the Earth sciences will occupy a choice position among the space sciences.

3. RBV and MSS ('ERTS 1' and 'B')

The first two satellites reserved for Earth sciences (of the Nimbus type), orbit at an altitude slightly higher than 911 km. The satellite crosses the equator at 9:30 in a

* This is a real 'break-through' in remote sensing. I think we must recall that the scientist responsible is Dr W. Hovis at Goddard Space Flight Center, Code 652.

southbound direction. The west-east band measures only 180 km (over 3000 km with the 'classic' Nimbus satellites). Two adjacent bands scanned by the space vessel are separated by 13 orbits. As an example, when the New York City region is in view in orbit No. x, the Pittsburgh region is scanned two days later, the orbit number being $x+26$.

The hour window used is not very opportune. At 45° north latitude on a descending (southbound) trajectory, the local solar time varies between 10:00 and 10:15, while the best spectral quality is obtained very close to midday. These time differences become catastrophic when we consider the ascending (northbound) trajectory: in this case, it would be between 20:30 and 20:45, much too soon after sundown (see Figure 11 for time differences). We know that the equipment on board ERTS 1 operates only in the daytime (visible and near infrared); on the other hand, in ERTS B, channel 5 of the multispectral spectrometer works equally well day and night (10.4–12.6 μm).

3.1. RBV television system (Return Beam Vidicon camera)

This instrument, constructed for very specific purposes, has a number of advantages over previous instruments; in particular, it has high resolution and greater sensitivity. It has three TV cameras:

Camera No. 1: resolution – 4500 lines, spectral region – 475–574 μm (green);
Camera No. 2: resolution – 4500 lines; spectral region – 580–680 μm (yellow-orange);
Camera No. 3: resolution – 3400 lines; spectral region – 690–830 μm (red and near infrared).

On the ground, we have available three individual images or a single image in false colors, derived from the three pictures received from the satellite.

3.2. MSS (Multispectral Spectrometer)

The spectrometer operates within four bands (ERTS 1) or five bands (ERTS B): channel 1, 500–600 mμ; channel 2, 600–700 mμ; channel 3, 700–800 mμ; channel 4, 0.8–1.1 μm and channel 5, 10.4–12.6 μm, the only channel detecting emitted infrared.

Spectral bands 1 to 4 benefit from ground resolution of 100 m and channel no. 5, a little over 200 m. It is hardly necessary to stress the importance of this multispectral spectrometer. However, I would like to draw attention to channel 5 (ERTS B, 1975?) which transmits a thermal image of the regions overflown both day and night in an east-west band 180 km wide. The comparative study of the five spectral bands (four of them measure daytime reflectivity; and the fifth, emitted infrared) permits identification of most terrestrial formations on a scale that allows detailed studies to be made. This leads to 'results' that are not only technically satisfactory, but are comparable to those obtained by traditional methods.

4. Hyper Frequencies and Thermal Infrared

The passive microwaves exhibit their usefulness with satellites comparable to the NASA-ESRO Spacelab. However, they can be fruitfully used on other platforms, like aircraft, or in the field (ground truth sites). One point, related to the proper wavelength, deserves to be stressed. A wavelength that is too long could be difficult to 'handle', because of the antenna size and because of the growing galactic noise. With a shorter wavelength the antenna problem fades away, but, atmospheric obstacles might interfere (trouble provoked by rain if the wavelength is not long enough, absorption by molecular oxygen and water vapor at wavelengths shorter than roughly 1.40–1.45 cm). NASA 'made up its mind', choosing a perfectly accurate frequency, i.e. 19.35 GHz (1.55 cm) and the author feels that such a good frequency should be permanently retained.

As stated earlier, such a radiometer combined with a 'classical' thermal infrared's, would permit us to instantly derive the most characteristic feature of any terrestrial object, namely its emissivity in the milli-centimetric wavelengths.

5. Processing of Information Received from Earth Resources Satellites

I will set aside the case of Nimbus 5 and F; their methods of operation are the same as for the earlier Nimbus (see Chapter IV and Appendix D). For ERTS, the situation is different.

For easily understood reasons, Earth coverage cannot be total as with Nimbus 1 through 5, since the instruments on board cannot be made to work continuously. Consequently, the ERTS satellites work only for a relatively short time. At the initial stage, only the territory of North America and selected foreign countries are covered, the satellite being 'inert' while overflying the remainder of the globe. The data acquisition centers in direct line of sight are found in Maryland, North Carolina, Texas, and Alaska. Other possible sites are being examined, for instance, a center at Port Churchill (Canada) which could take over from Greenbelt, Maryland. A station in Western Germany is also being planned. Finally, according to user demand, centers might be installed in any other part of the Earth, such as Spain, South Africa, South America, Madagascar, or Australia.

The processing operations can be divided into three main phases:

(1) The signals received from the two instruments are first demodulated, then recorded on magnetic tape.

(2) From the magnetic tape prepared at the receiving station, a new series of tapes is made, then films on which the geographical coordinates are overprinted. 'Pictures' are then taken from these and sent to the users (see Appendix E). This work is done systematically, and there is normally an 18-day gap between information acquisition and distribution.

For the United States alone, 45 sets are made up, each set consisting of three TV images and four (or five) spectrometer images. Each of these seven (or eight) images covers an area of 180×180 km.

(3) The images may be in either black and white or color, as may be required. Part of the information, especially for channel 5 (ERTS B, 1975) of the multispectral spectrometer can be digitized on demand. The maps prepared with the aid of ERTS information can be on a scale such as 1:25000, at least for the draft map. In my opinion, the final map should not be on a scale any larger than 1:100000 due to the problem of resolution.

There is a pressing need to digitize channel 5, and the same goes for channel 4. In the visible (channels 1 and 3 and television), the images supplied by NASA can be studied directly, as distortions are ironed out while the information is being processed at Goddard Space Flight Center. However, it is strongly advisable:

(1) to use the multispectral projection system for the three TV photographs (see Figure 4);

(2) to work from numerical values obtained by a densitometer procedure or retain the effective radiance values; in a pinch, to transform values into reflectance or radiance temperatures: the only way to avoid making approximations and to end up with valid results.

CHAPTER IV

PROCESSING OF INFORMATION RECEIVED FROM METEOROLOGICAL SATELLITES

There are five aspects to be examined: (1) to note what is happening on board the satellites; (2) production of photographic records; (3) preparation of the digitized magnetic tape; (4) preparation of digitized maps; and (5) cartographic problems.

1. Onboard Detection and Retransmission to Earth

1.1. Television

The TV camera of the IDCS system (Nimbus 3 and 4) has a wide-angle lens (103°); the focal length is 5.7 mm. On the ground, resolution is 3.7 km. Each image is composed of 800 lines and scanning is done in 3′20″ or one-quarter second per line, of which 0.225 s is for the active phase (signal reception) and 0.025 s for the passive phase.

The signals are recorded on board on one of the five tracks of one of the two tape-recorders, and retransmitted to Earth in the S band frequencies (1.7025 GHz; 17.61 cm).

1.2. Infrared

We will take the case of THIR on Nimbus 4. You will remember that the radiometer is operating in two spectral bands, 10.5–12.5 μm (atmospheric window) and 6.7 μm (water vapor band) (see Plate D, photo 1, opposite page 81). The latter channel measures the humidity of the upper atmosphere and stratosphere; through it the jet stream can be detected, and displacements of the front systems tracked. The ground resolution of this channel is 27 km.

An optical system distributes the radiation received on the water vapor and atmospheric window detectors. The radiations are amplified and transformed into electric signals (voltage), and recorded onto two of the five tracks of the HDRSS system. When interrogated, the magnetic tapes play back and transmit to Earth the information stored on the magnetic tape, along with data from other experiments assigned to the same equipment, with the time code.

The APT stations are simple and inexpensive to build and can receive when the satellite is in direct line of sight without the information being prerecorded on the HDRSS system. The NASA department Nimbus APT, code 450, Goddard Space Flight Center, Greenbelt, Maryland 20771, provides full information on building and operating these receivers. I have borrowed the details in Table XIV, below, from a circular letter dated November 26, 1969, from the coordinator of this service, John

TABLE XIV

Voltage-radiation temperature correlations (for an APT receiver, Nimbus 3 nocturnal infrared in the spectral band 3.5–4.2 m)

Voltage (DC)	Temperature of detector on satellite, in °C							
	−77.5	−75.0	−72.5	−70.0	−67.5	−65.0	−62.5	−60.0
	(Radiance temperatures in degrees Kelvin)							
0 to 0.3	190	190	190	190	190	190	190	190
0.4	205.0	205.5	205.8	206.1	206.4	206.7	206.9	207.1
0.5	209.6	210.3	210.8	211.2	211.6	211.9	212.3	212.6
1.0	224.5	225.8	226.8	227.7	228.5	229.3	230.0	230.5
2.0	243.3	245.6	247.3	248.9	250.3	251.4	252.5	253.5
3.0	260.7	263.9	266.4	268.7	270.6	272.3	273.9	275.4
4.0	282.3	286.8	290.5	293.7	296.5	299.0	301.3	303.4
4.2	288.6	293.5	297.5	301.1	304.1	306.8	309.4	311.7
4.4	295.6	300.9	305.4	309.2	312.5	315.6	318.3	321.0
4.6	302.4	308.3	313.1	317.3	321.0	324.3	327.4	330.4
5.0	315.6	322.4	328.1	333.1	337.4	341.3	345.0	348.5
6.0 (maximum)	337.3	346.0	353.1	359.5	365.7	372.0	378.6	385.6

Lindstrom: correlations between voltage (at the receiving station) and radiance temperatures (in degrees Kelvin), for different temperatures of the infrared cell on board the space vessel, in degrees centigrade (valid for Nimbus 3, radiometer, 3.5–4.2 μm). This direct readout data is transmitted on the 136.95 MHz (1.79 m) frequency.

2. Production of Photographic Records (Photofacsimile)

It has become customary to use the term *photofacsimile* although the machine which gave its name to the product is no longer in use.

At the receiving station, the signals are broken down channel by channel, recorded on magnetic tape, and transmitted to Goddard. Here, the frequency modulated signals are demodulated, synchronized, and recorded onto a magnetic tape known as *analog* or *raw tape*, which serves to produce two *formats*, i.e. two main types of documents: facsimile photographs and a digitized magnetic tape, NMRT (*N*imbus *M*eteorological *R*adiation *T*ape) essential for preparing digitized maps.

2.1. Graphic reproduction of photofacsimile

To reconvert the picture transmitted from the satellite, the analog tape controls a beam of light that scans photographic paper under control of the scanning conditions of the mirror on board the satellite. The photographic paper unwinds at a speed synchronized with that of the satellite. The intensity of the light beam is directly controlled electronically by the analog tape. Thus, a picture is reproduced, line by line, on 70 mm film.

As the sensitivity of the detector is 1 K, it would theoretically be possible to reproduce pictures incorporating all of the tonal shades recorded on board the space vessel. It does not work out this way in practice. Due to the number of spurious signals (static, noise) if every signal were to be reproduced on the picture, it would be distorted out of all recognition. Also, although the system I have just described has been improved by utilization of a cathode ray tube, the inertia factor has to be allowed for.

Despite the reduction at ten temperature levels that we will take a look at below, the *inertia factor* distorts the pictures especially when a thermal gradient, for example a land-sea boundary, extends in the flight direction of the satellite – a very roughly north-south oriented boundary. In this case, the inertia factor usually shows up as a narrow, light strip along the coastline. I point this out because errors have already been committed in the analysis of photofacsimile when some interpreters have discovered non-existent features.

In practice, ten thermal levels, are used (Nimbus 4) giving ten shading-off tones corresponding to the following equivalent blackbody temperatures: over 330 K, 330–326; 326–314; 314–302; 302–288; 288–273; 273–256; 256–234; 234–196, and finally temperatures under 196 K. For the water vapor channel, the limits go from 270 to 200 K. A voltage of 5.33, on the average, corresponds to 330 K in the 10.5–12.5 μm band. Consequently, the intensity of the beam of light is very high, the film is overexposed, and thus gives a very deep black tone. A voltage of 0.843 to 0.857 corresponds to 190 K; this low value gives low light intensity and, thus, very light shades. This is the origin of the arbitrary definition of a 'positive' which shows high temperatures by dark shades and low temperatures by light shades, or exactly the opposite of pictures produced by direct readout radiometers.

At Goddard, the first film obtained serves to make a negative (*master*) from which the records requested by users are produced; the original film is kept in safe storage.

2.2. Symbols appearing on photofacsimile

As the photographic emulsion is scanned line by line, the geographic coordinates are simultaneously over-printed from telemetric and time data that is transmitted at the same time as the electromagnetic signals detected by the instruments on board the satellite. This work is done manually, and the operator's lapses of attention may be seen by sudden breaks affecting meridians and parallels. These are drawn in as dashed lines spaced 2° apart for both latitude and longitude, but only coordinates 10° apart (0°, 10°, 20°, 30°, etc.) are plotted on the facsimile photos. Every 30 degrees of latitude, a cross is placed at the point immediately below the satellite position and the longitudinal and latitudinal coordinates (always 0°, 30°, and 60° N and S) are marked. The east or west longitude has been marked E or W since Nimbus 3; previously only the east longitude positions were marked, for example 312 which meant 48° W.

On the margins of the photographic strips, white dashes indicate time intervals of 2 min. At the bottom of each photographic strip appear the name of the satellite, channel, day or night (D or N), day of the year, and GMT time at the beginning of the orbit shown (the beginning being, by definition, the bottom of the strip even if the flight was southbound). We might thus read: NIMBUS IV ORBIT 503 THIR D 11.5 μ and 157074825. This means: D for day, channel 11.5 μm, 157th day of the year (6 June for non-leapyears), 7h48′25″, universal time (GMT).

2.3. Errors in location affecting photofacsimile

Despite every precaution, the coordinates are rarely just where they should be for a very simple reason: the satellite is assumed to be perfectly stabilized so that the points in the center of the photographic strip are considered to be directly below the satellite.

Actually, the vehicle pitches, rolls, and yaws slightly. These movements, fairly large with Nimbus 1, have been smoothed out little by little. With Nimbus 3 and 4, they did not exceed an angle of $\pm 1°$ – very small but enough so that errors in location of 19 km were recorded directly below the satellite. On the scale of photofacsimile this is scarcely more than the thickness of a penstroke, but becomes more important when we are working on real research records, i.e. digitized maps.

2.4. Utilization of photographic records

These facsimile photos, well known to the general public, cannot be considered as working tools for Earth sciences, while in meteorology, they can be used directly and have rendered invaluable service. The U.S. Weather Bureau (ESSA, Suitland, Maryland) has developed a procedure that automatically eliminates all distortions: one thus obtains composite photographs which, alone, give sufficiently precise information on the geographic location of the phenomena studied: distances, rate of movement of frontal systems, etc. Plotted photographs are not made at Goddard Space Flight Center.

It is not only for reasons of scale and distortion that these documents are poor servants of geographical research. The reproduction system of ten thermal levels (or levels of reflectivity for infrared albedo) is the real culprit. Even if the complete thermal range (from 190 to 330 K) could be reproduced, the eye could not distinguish more than 6 or 7 levels, depending on the individual. Thus, even under (rarely-occurring) optimal photographic reproduction conditions, heat differences smaller than 10 K cannot be appreciated.

Perhaps the true reason underlying the limited value of these records (when we move on from reading to research work) is, basically, defects inherent in the reproduction system itself. Several machines are used, each of which produces slightly different images. Some are more contrasted, others flatter, from the same magnetic tape. We saw a little while back that thermal gradients with an approximately N-S orientation can lead to confusion by introducing nonexistent features (inertia factor). This is not the end of the story: heat contrasts that are actually quite small may show up sharply contrasted. The inverse is just as true. When the digitized maps are finally prepared, these nonexistent contrasts disappear, and I have often heard beginning interpreters complain of this: they prefer photographs because they are ‘clearer’ than the maps, while actually the maps are telling the truth and the photofacsimile are wrong.

Photographs do, however, have some important practical uses. First of all, they are the only means we have available to plot the most suitable orbits: we can find out immediately whether or not clouds prohibit us from working with the orbit. The cost price of digitized maps is so high that no-one can afford the luxury of systematically programming every orbit.

Finally, these photographs are irreplaceable for teaching purposes: the most important function of all.

3. Digitized Magnetic Tapes (NMRT)

Research cannot be properly carried out without the aid of digitized maps. These are obtained first by calibrating and digitizing the analog magnetic tape, producing the digital magnetic tape which alone can be fed into the IBM 360-91 computer currently used at Goddard. This magnetic tape is known as NMRT (*N*imbus *M*eteorological *R*adiation *T*ape) to which is added either HRIR or MRIR for Nimbus 1, 2, and 3, THIR, for Nimbus 4, SCMR for Nimbus 5, F, etc.

3.1. Problem of Interference

The preparation of digitized tapes has numerous obstacles to overcome. A CDC 924 computer is used for calibration; then, combining the calibrated tape with the telemetric information, we obtain the definitive digitized tape (NMRT). This is a huge and costly job. Calibration itself is rarely correct as far as absolute values go, but is valuable if we are wise enough to consider only the reciprocal relationships, the gradients.

The *signal-to-noise ratio* must be favorable enough so that the background noise will not distort the final record too severely. Among these background noises are normal noises and random noises. To make myself clear, I will use the comparison of the family TV screen. When the set is properly adjusted, the antenna correctly pointed, and the picture contrasted and clear, small details may be distinguished. In this case the signal-to-noise ratio is high. If adjustment is incorrect we can get ghost images, echoes, distortions, snow, etc. In this case the signal-to-noise ratio is low but can be improved since the family set is within easy reach. This is clearly not the case when the knobs to be turned are on a satellite.

Random noise can be temporary, as that caused on a TV set by a passing automobile with a defective antistatic system. Correction is simple. Noise can also be permanent, caused by an irreparable malfunction in the receiving set (in which case we pick up another one at the store) or in the transmitter. To come back to satellites, it goes without saying that such a fault, accidental though it may be, is there to stay. This is what happened on board Nimbus 2 through 5. On the other hand, when the cause of the trouble has been diagnosed, it can be remedied by the apparatus described below.

For example, the oblique lines affecting Nimbus 1 photofacsimile came from periodic interference having a frequency of 16 Hz. This was corrected for later satellites, which are affected by medium frequency interference (200 Hz). In the latter case, errors reached and exceeded 3 K upon calibration, and increased considerably at low temperatures (Williamson, 1970).

If interference introduced during ground operations is (or could be) eliminated, it is not the case for interference originating on the satellite. In this case, digital filters are applied which attenuate to a degree the area occupied by undesirable signals.

3.2. Contrast between Successful Injection into Orbit and Delays in Using Information

The problems involved in digitized map production began as far back as the first Nimbus and have not been resolved at the time of the fifth. Indeed, there is little likelihood that they will be in the near future, unless a solution is sought by the agencies associated with NASA or, better still, by an international scientific body.

The photofacsimile is produced as soon as the signals are received: in the next 48 h researchers can obtain the records of interest to them. For research reasons, the same time lag should apply to production of digitized maps, but in fact, the wait is a year or more. This has serious effects on fundamental research.

A great deal of time and skilled personnel are needed to determine the calibration system and to study interference characteristics. Several parts of these operations, especially the interference aspect, could be done in minimum time, a day or two after transmission, if we had work routines (which exist), skilled personnel (who exist) in sufficient numbers (which is very far from being the case).

Calibration, even rough calibration, could be accepted. This operation is enormously time-consuming because 'they' want absolute precision, which is not only impossible but quite purposeless. Unfortunately, many 'interpreters' are still convinced that the actual temperatures of the land or sea surfaces can be found using blackbody radiation values plus the necessary 'corrections'. This is chasing after an impossible dream. It is wasting both the time of the calibrators and of those users who would be content with distributions, gradients, thermal values, or reflectivity percentages. When this has been resolved, perhaps we could get digitized maps inside reasonable time lags, i.e., two to three weeks at most, after signal reception.

Many of my colleagues, some NASA people and some from private organizations, believe as I do that the contrast between successful launching of a satellite and the scientific processing capability is so disproportionately huge that it looks like an unprecedented squandering of public resources.

In 1971, nine-tenths of the Nimbus-2 orbits (for high resolution) were processed. The Nimbus-1 NMRT tapes have mysteriously disappeared. Less than 10% of the orbits of Nimbus 3 have been calibrated and reduced to figures. In June 1971, of more than 5000 orbits flown by Nimbus 4, a tiny number (80? 100?) has been reduced to figures, and one still has to go directly to these NMRT tapes to obtain the data necessary for research.

This is the most striking weakness in all meteorological satellite programs. I believe it is critical to focus attention on these problems and to seek a solution that will only be possible with international cooperation.

4. Digitized Maps

As soon as the digitized numerical tape is ready, the difficulties disappear and computer-produced digitized maps can be obtained in 24 h. When they have the necessary information (see Appendix D) the IBM 360-91 operators follow user requests as to scale, type of symbol for areas of equal value, latitudinal and longitudinal extension, etc.

Except for the Arctic and Antarctic regions, the Mercator projection is routinely used. The coordinates are shown by a code keyed at the end of the map; but, as we have already seen, it is rare for them to be correctly placed. Empirical methods enable the parallels and meridians to be placed fairly correctly: a correction table is given at the end of Appendix D.

4.1. Maps with temperature or reflectance values

Maps drawn to the scale of 1:1000000 have eight points numbered per degree of longitude, i.e. every 12 to 14 km, while resolution is about 8 km in the center. This means that each value is the average of several signals, a fact that sets off the influence of spurious signals. A population map, in the statistical sense of the term, indicates point by point the number of signals whose average gives the printed number. The number of signals varies between 1 to 10 or 12, in general; on the 1:1000000 scale, from 4 to 5. When a single signal (population: 1) has served as a basis for the thermal or reflectance value, it is not advisable to use the number printed by the computer since the influence of spurious signals has not been narrowed down. When the interference is greater than the signal (excessive voltage) the computer is programmed to ignore the fact, and leaves a blank, A few years ago, the machine was programmed to write in the general mean of the map, which caused no little confusion.

4.2. Maps without isolines

These are the simplest of all. They come in the form of grids of small crosses spaced at half-inch intervals accompanied by a four-figure number. The work of the analyst is simple but lengthy and tiresome: he draws in the lines of equal reflectance or equal temperature. Actually, the work is often done incorrectly.

4.3. Distinguishing equal-value areas ('contouring')

The above work can be done in part mechanically, setting off areas where temperature (or reflectance) is between A and B degrees or percent. There is no limit as to the choices of intervals – every 10°, 5°, 2°, etc.

Meteorologists have adopted 10° in most cases. After a number of trials, I worked out a routine for 2°, as experiments with 1° were more often disappointing than successful. Programming at 5°, for example, gives nothing better than the facsimile photos can offer as most of the detail is lost.

The above applies to the scale of 1:1 000 000 as far as possible. It will easily be understood that, even using 1°, reduction to scale blurs out most of the detail while reducing the influence of spurious signals.

In practice, matters are very simple: we assign letters to temperatures, for example A for 280 and 281 K, a blank for 282 and 283 K, then B for 284 and 285 K, etc. In 1967, I stopped at H (giving 16 thermal levels). However, when the range exceeded 16 thermal or reflectance levels, the same series of letters was used for values much higher or lower. In 1969, we succeeded, not without some difficulty, in persuading the computer operators to modify the format so that we could use the whole alphabet from A to Z, thus distinguishing 52 levels which was more than enough. For the heat values, the user must indicate the 'height' at which he wants the procedure to be employed. Oceanographers set their 'beginning' or 'end' in order to show up specific areas only on oceans. We do exactly the opposite. Oceanographers and earth science specialists arrange for the clouds scattered here and there to be left blank, while meteorologists deal with much greater temperature or reflectance ranges.

One example will make things clearer. When I was studying a region overflown in the daytime (about midday) in the summer, I asked for a temperature level beginning at around 290 K in order to leave the oceans without reference letters. In this case, a blank would be left for values equal to and less than 291 K; letter A would indicate temperatures of 292–293 K, then a blank for 294–295 K, B for 296–297 K, etc.

Experience has shown that the computer does a more thorough job than work done with raw values. This demands an explanation. The values stored on the magnetic tape (NMRT) always have a decimal. The computer does not round off: it leaves out the decimal. Instead of printing 292 K for 291.9, it prints 291 K. On the other hand, when it is programmed to set off areas having equal temperature, the computer makes the differentiation. Imagine an NMRT tape with the figures 280.3 and 281.9 K. On the 'rough' map we read 280 and 281 K, but the differentiated area system will print out the letter C, for instance, for the first number and a blank for the second, rounded off correctly to 282 K.

The situation is similar for the infrared albedo expressed in tenths of a percent. Thus, the number printed by the computer, 0235, means 23.5%. In June–July 1969, after some trial and error, I managed to make a definitive choice of numbers on these numbers with one decimal (which, in technical jargon, we call the result of multiplication, the multiplier in this case being 10 000), the base being 10 and the interval being −10. In simpler terms, I program at 1%, and a peripheral storage is permanently assigned to this type of work (multiplier and base, the user being free to request intervals of 2% in which case we would have −20, or 0.1% giving −01, etc.).

Figure 12 is an example of a map set up in this way. It is for a high resolution radiometer operating on the Nimbus 3 satellite in the daytime. Orbit no. 711, 6 June, 1969 was selected because of the high quality of reflected infrared, spectral band 0.7–1.3 μm. A routine some months old was used to convert the effective radiation

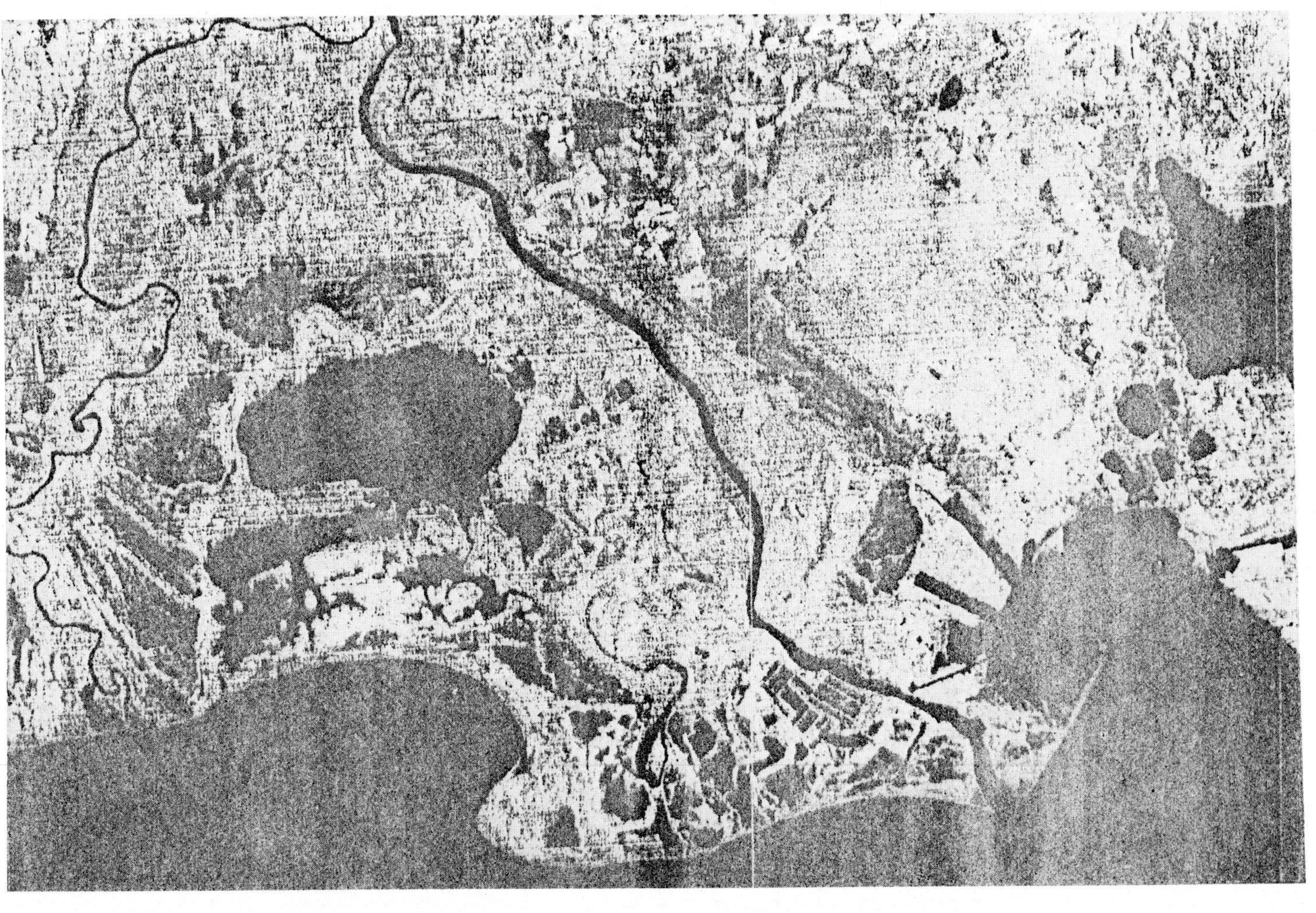

Plate C. The Rhone delta seen by ERTS 1, channel 7 (MSS channel 4, reflected infrared).

into albedo (reflectance) and the percentages are shown with one decimal, setting off equal value areas at 1%. Running your eye down the line of C's, you will have no difficulty in recognizing part of Saudi Arabia. On the left, the numbers 001, 002, etc. indicate the *latitude*: 001 = 12° N, 009 = 12.9° N, 016 = 13.8° N. *Longitude* is shown by the numbers printed along the bottom; 033 = 316° W (44° E); 026 = 316.9° W (43.1° E), etc. (Eight crosses per degree of longitude.) A conversion table is provided with each map.

The letters represent: A = 0 to 0.9%; B = 2 to 2.9%; C = 4 to 4.9%, the blanks represent intermediate values 1–1.9%, 3–3.9%, etc.

It will immediately be noted that the latitude position 12° N is incorrect. The 13° parallel (between 009 and 010) seems to be in the proper place as well as the 44° E meridian, but there would be little point in being more precise.

4.4. 'False-color' maps

The same NMRT magnetic tape is also used to translate temperatures or percentages into colors according to an arbitrary system defined by the total number of shades (50 maximum) and their combination. The results are extremely satisfactory but many people think the lines showing equal level is a more practical procedure, though it takes considerable time to draw up the cartogram supplied by the computer. A computer has also been programmed to plot the isolines directly. The results are good, nothing more, and very inferior to those obtained by the method described above.

5. Cartographic Problems

These problems have been dealt with in greater or lesser detail in the preceding pages. They can be divided into three categories, relating to: (1) spurious signals, (2) satellite movements; and (3) the problem of scale, this being the most important of all (Pouquet, 1972, Montreal).

Plate C.

The Rhone delta seen by ERTS 1 (ERTS 1 channel 0.8–1.2 μm). In the reflected infrared, water appears black (sea, lakes, rivers). The Grand Rhône (Great Rhone), and the Petit Rhône (Little Rhone) are striking with their very black (thick and thin) lines. The Etang de Berre is partly seen (right side) as is the Etang de Vaccarès (between the two branches of the river). The new industrial complex of Fos is clearly depicted, just East of the lower Grand Rhône.

The curved stripes of the western Delta correspond to successive positions of the same delta since the termination of the last glacial period. The former general shorelines are easy to see as well as ancient river courses. It is even possible to read, 'as an open book', the recent dynamism of this evolution, opposing the western delta, perfectly visible to the eastern section where the continuity is not affected by the 'cut offs' seen along the present day shoreline of the part built by the Petit Rhône. The huge amount of mud carried by the Grand Rhône is shown by the growing eastern delta.

This is the first time it is possible to read the geomorphological evolution of that part of Southern France, with many new faccts. Needless to say, this photograph has been widely published.

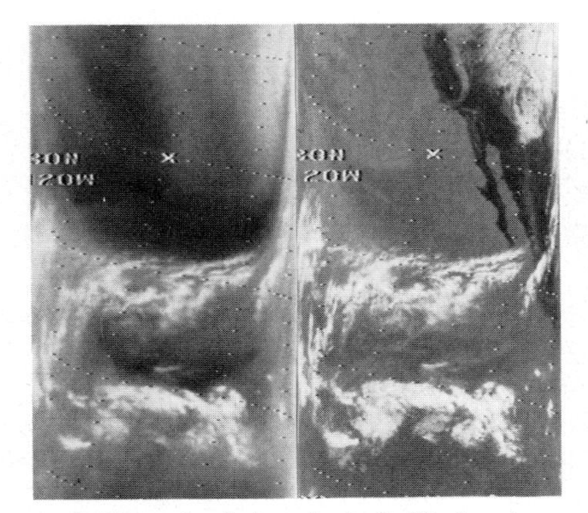

1. Example of photofacsimile Nimbus 4.

2. The Earth seen from ATS 1 satellite.

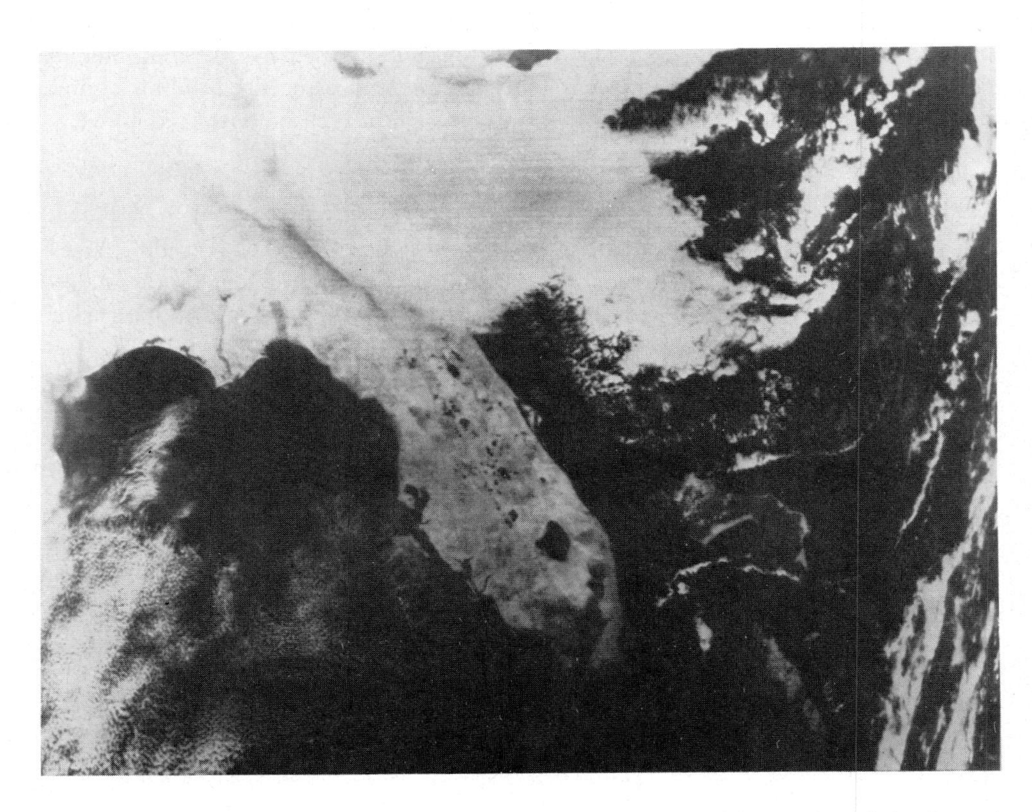

3. Florida seen by Nimbus 5, SCMR nighttime thermal infrared, very high ground resolution (around 600 m), on 24 December 1972 (Orbit 173).

5.1. Spurious signal problems

Spurious signals transmitted by the instruments are almost always geometrically arranged: they are found in bands perpendicular to the flight axis and, very often, in more or less close-packed bands shaped like an arc of a circle. Figure 13 gives an example that might be of little significance. The spurious signals are in a fringe perpendicular to the flight axis, marked 'noisy area' on the map. I have actually completed the computer's work by joining up the isolines on each side of the noisy band. The orbit treated in this way offers not only an example of spurious noise but two additional points of interest. First, one can observe the zone of influence of Lake Chad towards the east and south-east, and note the sharp image of the Termet massif. This stands out not because of height differences, since altitude does not affect the reflected near infrared, but because the rock material differs from that on the periphery. The second reason is the late date on which the radiations were recorded – late November, 1969 – at which point the onboard taperecorders were becoming more and more defective. We can see that, despite this deficiency in recording, the signals received from the satellite can still be used, noise notwithstanding.

The map in Figure 14 illustrates another case. This record, prepared from an excellent orbit, shows some features of West Africa, especially the SW-NE oriented physiographic region where the community of Atar occupies a remarkable position, at the intersection of two identical features. However, I want to draw your attention to the triangular-shaped region. One of the sides of this triangle is almost ruler-straight. The north side is more undulating; and the base coincides with the shoreline. In the middle of the triangle we note the preponderance of circular or elliptical

---

Plate D.

1. Example of photofacsimile, Nimbus 4, temperature moisture radiometer, daytime values. Orbit 302, 30 April 1970.

On the left, water vapor channel (6.7 μm). On the right, 10.5–12.5 μm channel.

The infrared emitted by the clouds in the two spectral bands gives similar pictures. On the other hand high atmospheric humidity (light shades) or low humidity (dark shades) is quite evident on the left photograph while it scarcely affects that on the right.

2. The Earth seen by ATS 1 satellite, altitude 30 000–35 000 km. Meteorological situation over the Pacific. Note the high pressure region off the coast of Baja California. The tropical convergence zone is particularly clear.

3. This is a 'unique' picture: soon after, the splendid radiometer designed by Dr W. Hovis at Goddard Space Flight Center, went out of order. What we got is promising for the future: one can see how many facts are brightly depicted, thanks to the very fine resolution of the radiometer (spatial resolution of 0.6 mrad). With this photograph, a "page has been turned" in the history of Remote Sensing.

Looking only at the peninsula, compare Cape Kennedy, Tampa Bay, Charlotte Harbor, the whole shoreline between Cape Romano and Cape Sable... with existing large-scale maps and small changes are easy to pin-point. The same can be said of the Everglades area.

However, the most striking fact is shown by the almost straight 'line', dividing the peninsula in two halves. This line is emphasized by numerous tiny ponds or lakes. What is its significance? Tectonic accident probably. Anyway, Dr Hovis' radiometer picked up a very interesting feature which, if we could draw a digitized map, would be much easier to study.

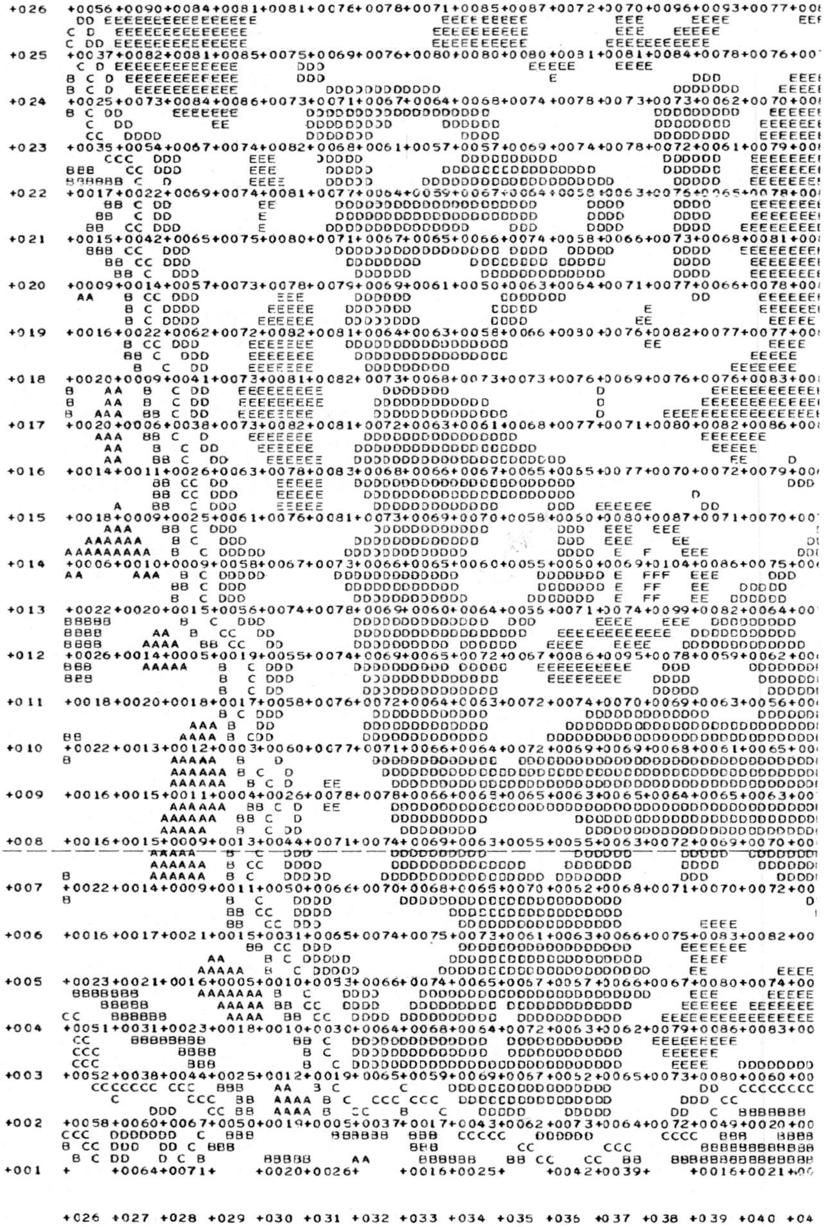

```
+026  +0056+0090+0084+0081+0081+0076+0078+0071+0085+0087+0072+0070+0096+0093+0077+00
+025  +0037+0082+0081+0085+0075+0069+0076+0080+0080+0080+0081+0081+0084+0078+0076+00
+024  +0025+0073+0084+0086+0073+0071+0067+0064+0068+0074+0078+0073+0073+0062+0070+00
+023  +0035+0054+0067+0074+0082+0068+0061+0057+0057+0069+0074+0078+0072+0061+0079+00
+022  +0017+0022+0069+0074+0081+0077+0064+0059+0067+0064+0058+0063+0076+0065+0078+00
+021  +0015+0042+0065+0075+0080+0071+0067+0065+0066+0074+0058+0066+0073+0068+0081+00
+020  +0009+0014+0057+0073+0078+0079+0069+0061+0050+0063+0064+0071+0077+0066+0078+00
+019  +0016+0022+0062+0072+0082+0081+0064+0063+0058+0066+0080+0076+0082+0077+0077+00
+018  +0020+0009+0041+0073+0081+0082+0073+0068+0073+0073+0076+0069+0076+0076+0083+00
+017  +0020+0006+0038+0073+0082+0081+0072+0063+0061+0068+0077+0071+0080+0082+0086+00
+016  +0014+0011+0026+0063+0078+0083+0068+0066+0067+0065+0055+0077+0070+0072+0079+00
+015  +0018+0009+0025+0061+0076+0081+0073+0069+0070+0058+0050+0080+0087+0071+0070+00
+014  +0006+0010+0009+0058+0067+0073+0066+0065+0060+0055+0050+0069+0104+0086+0075+00
+013  +0022+0020+0015+0056+0074+0078+0069+0060+0064+0056+0071+0074+0099+0082+0064+00
+012  +0026+0014+0005+0019+0055+0074+0069+0065+0072+0067+0086+0095+0078+0059+0062+00
+011  +0018+0020+0018+0017+0058+0076+0072+0064+0063+0072+0074+0070+0069+0063+0056+00
+010  +0022+0013+0012+0003+0060+0077+0071+0066+0064+0072+0069+0069+0068+0061+0065+00
+009  +0016+0015+0011+0004+0026+0078+0078+0066+0065+0065+0063+0065+0064+0065+0063+00
+008  +0016+0015+0009+0013+0044+0071+0074+0069+0063+0055+0055+0063+0072+0069+0070+00
+007  +0022+0014+0009+0011+0050+0066+0070+0068+0065+0070+0062+0068+0071+0070+0072+00
+006  +0016+0017+0021+0015+0031+0065+0074+0075+0073+0061+0063+0066+0075+0083+0082+00
+005  +0023+0021+0016+0005+0010+0053+0066+0074+0065+0067+0057+0066+0067+0080+0074+00
+004  +0051+0031+0023+0018+0010+0030+0064+0068+0064+0072+0063+0062+0079+0086+0083+00
+003  +0052+0038+0044+0025+0012+0019+0065+0059+0069+0067+0052+0065+0073+0080+0060+00
+002  +0058+0060+0067+0050+0019+0005+0037+0017+0043+0062+0073+0064+0072+0049+0020+00
+001  +     +0064+0071+     +0020+0026+     +0016+0025+     +0042+0039+     +0016+0021+00

      +026 +027 +028 +029 +030 +031 +032 +033 +034 +035 +036 +037 +038 +039 +040 +04
```

Fig. 12. Computer-produced outline of equal-value areas (contouring). Nimbus 3, orbit 711; infrared albedo in the 0.7–1.3 μ band. – Left column shows latitude; bottom row of figure shows longitude (see text).

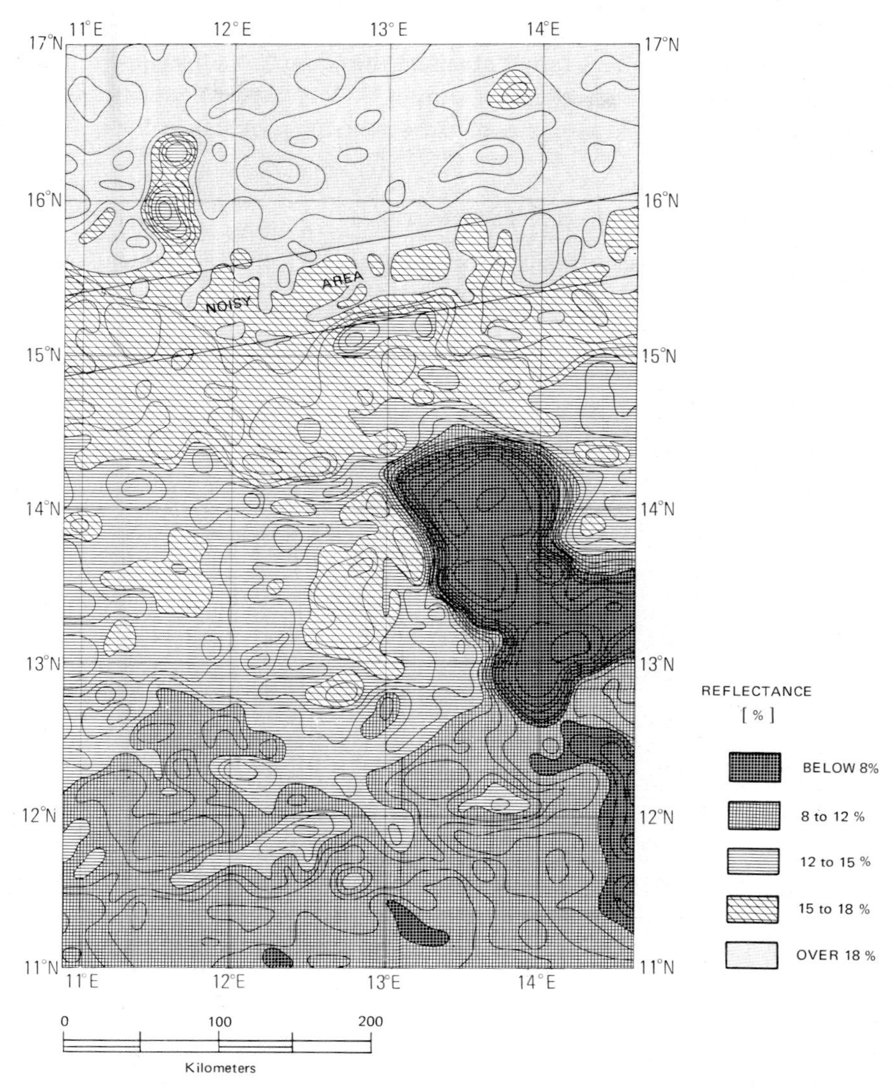

Fig. 13. Examples of non-troublesome noise. Lake Chad, central and western region. – Nimbus 3, infrared albedo (in %, band 0.7–1.3 μm), orbit 3045, 27 November 1969. – Key (starting with darkest shade): reflectance less than 8%; 8–12%; 12–15% (fine horizontal lines); 15–18%; over 18%. The noisy area is so marked.

shapes; below the southern border, however, these circles or ellipses tend to stretch rightward, parallel to this side of the triangle. On the other hand, we distinguish no geometric anomalies along the base, nor along the northern border.

This phenomenon turned up several times with maps of this and other regions. Our investigations to determine the real cause have been unsuccessful. We feel that geometrical shapes inside a triangle like this one are more than improbable, but we can find no logical explanation. We fell back on blaming the 'noise', but is this the answer? I hardly think so, and rather attribute the cause to an instrument malfunction between calibration and computer output*. Only experience will teach us to distrust such distribution of geometrical areas sometimes mistaken for natural features.

5.2. Problems introduced by satellite motion

Figure 14 presents a classic example of satellite yaw. This gives an angular shift suffi-

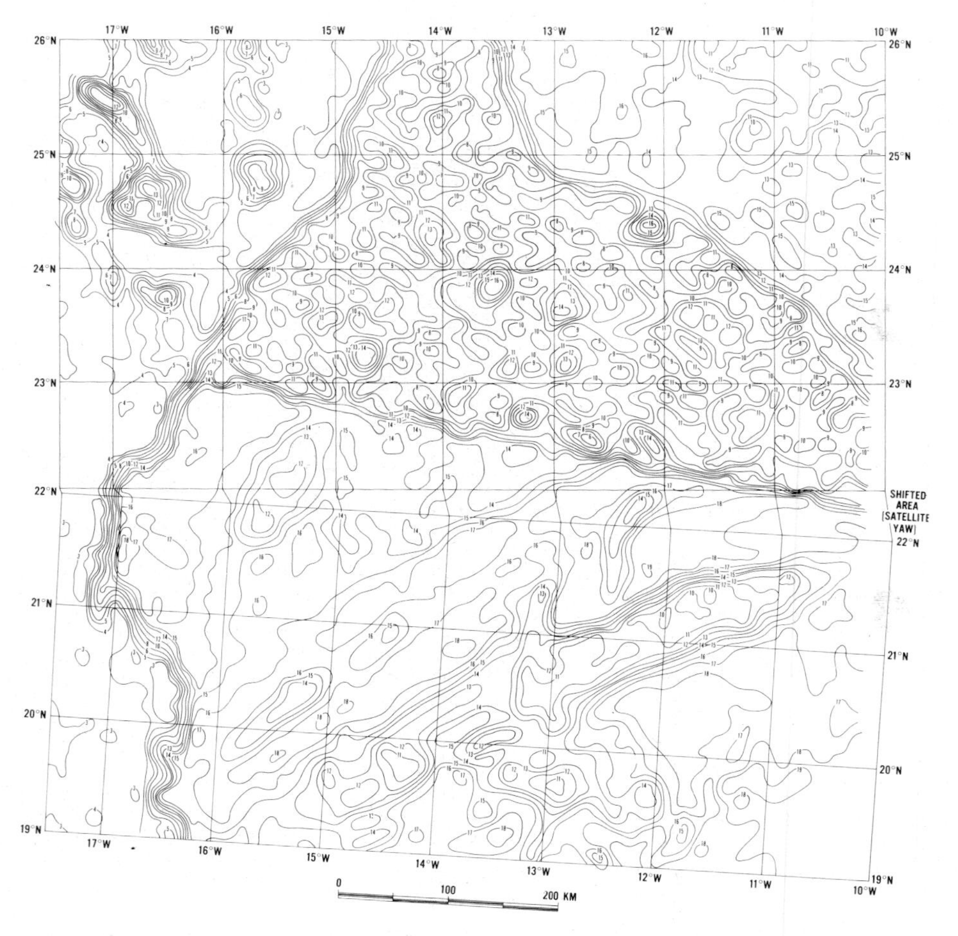

Fig. 14. Influence of noise and influence of satellite movement. Western Africa. – Infrared albedo in %; orbit 271, 4 May, 1969 (Nimbus 3). Region affected by noise: set off by sharp gradients in the south (almost linear) and in the north. Satellite yaw shows up by a shift in linearity of the coordinates.

* The same digitized magnetic tape (NMRT) once produced an excellent map and a second time, programmed for a different interval, a cartogram with the defect mentioned. Upon examining the phenomenon, I have compared the stretched line to 'weeping', very well known to those who have played a magnetic tape over a dirty head.

cient to cause an error in placement of the coordinates. In this case, for clarity of demonstration, I have kept the distribution of the values and shifted the coordinates to compensate for linear errors.

An example of satellite pitch is furnished by the map of Figure 15. Here again, linear errors have been compensated for by shifting the geographical coordinates. Here the pitch error is fairly small and could have been overlooked without introducing troublesome defects (remember the scale). On this same figure, I would like to point out the outcroppings and the scree at the foot of the Tidijka-Oualata cliffs.

The errors are committed by the computer which locates the areas successively scanned by the radiometer, placing the center point half-way between the right and left horizons. Consequently, an error by the satellite in accurately determining one horizon will result in displacement of the actual center.

From horizon to horizon, the strip scanned consists of an average of 400 individual *shots* out of a total of 1000. Assuming that the error in determining one of these horizons is in one shot, the angular deviation is thus $360°/(1000 \times 1.25) = 0.3°$ per shot. Reminder: the scanning mirror revolves at 48 rpm or 360° in 1.25 s.

Translated into ground resolution, this simple error in determining one of the two horizons causes a displacement of 5.4 km vertically below the satellite, and much more on the edges.

This phenomenon has another consequence, logical though odd, which often throws off beginners and even more experienced people. These localization errors are exaggerated when thermal gradients coincide with shorelines more or less perpendicular to the axis of the areas successively scanned, thus approximately N-S. In this situation, when one of the two horizons has been incorrectly recognized by the satellite (which is often the case), rectilinear shorelines become rather wavy and horizontal displacements can reach 10 to 20 km.

The cause of these errors is well known. The satellite stabilizes itself by looking at the right and left horizons. When a large cloud mass is on one of the horizons, the instrument on board mistakes the top of the clouds for the real horizon, causing an automatic attitude correction, i.e. a wobble.

When the maps are prepared, empirical corrections are applied. We use obvious landmarks such as river, and shorelines, and recalculate the correct position of the coordinates. It would be possible to correct mathematically by taking each sample one by one: 400 samples per line, and thousands of lines for a fairly small surface (one line is the same width as the resolution). It is my belief that on a scale as small as 1:1 000 000, this labor is out of proportion to the insignificant corrections it would give.

5.3. The Problem of Scale

Since ground resolution is about 8 km, the largest scale is that of 1:1 000 000, and even this involves some mental gymnastics. On the NMRT, the signals are separated. However, there is overlapping not only from orbit to orbit but also from line to line.

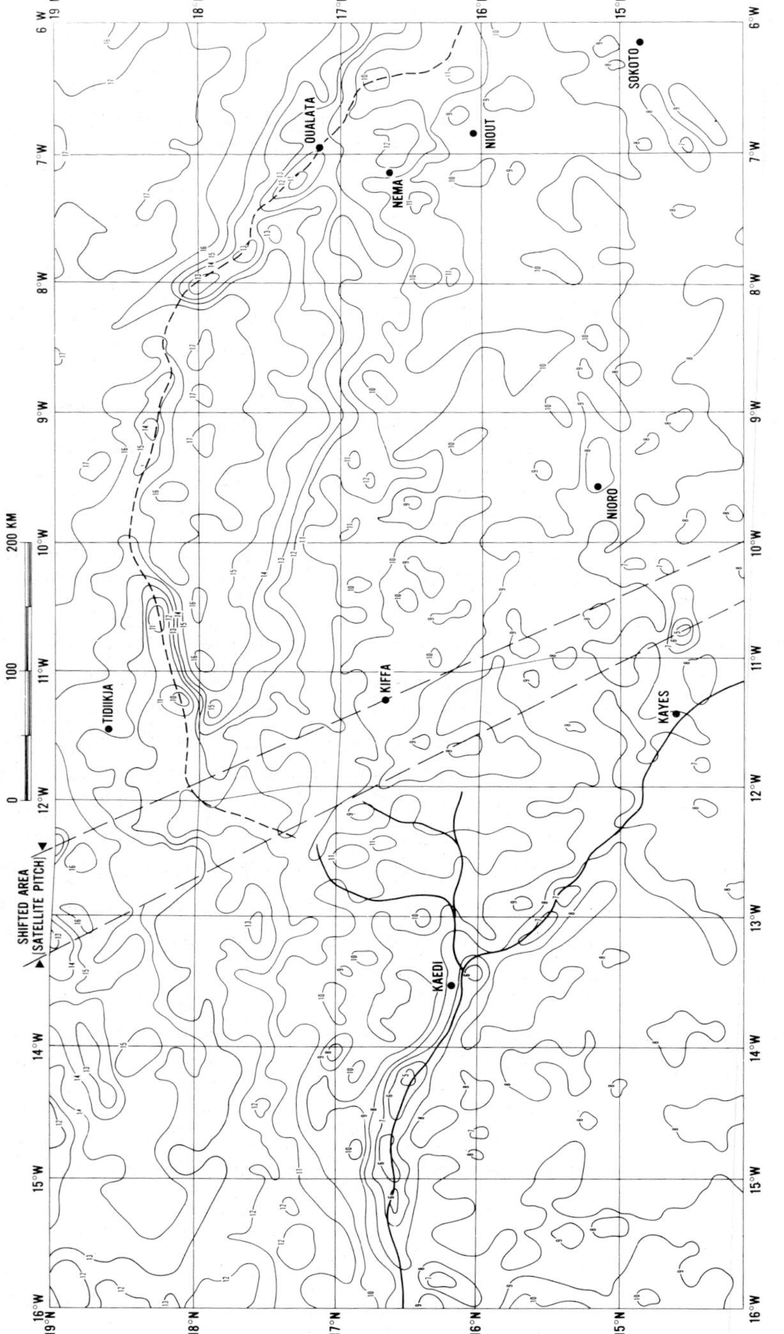

Fig. 15. Satellite pitch. Nimbus 3, orbit 2751, 5 November, 1969. Infrared albedo in %. The dashed lines give an approximate indication of the region most affected by satellite movement.

Thus, for a given ground area 8×8 km, we have not just one signal, as in theory we should have, but two, three, or more.

The routine for producing the maps is based on the use of sheets of paper with crosses every half inch. A simple calculation shows that there are eight crosses per degree of longitude on the 1:1000000 scale, 16 on the 1:500000 scale, 4 for 1:2000000, etc. This leads us to examine two aspects of the question at issue.

(a) *Separation of every signal.* For one degree of longitude we have a choice between 8 or 16 signals, to get the largest scale possible. Take a degree of longitude with an average value (at low latitudes) of 110 km. We know that the satellite distinguishes units about 8 to 10 km long when it is vertically above them, thus, in the most favorable case, 13 to 14 shots per degree instead of 16 as in the case of 1:500000. As a result, at least two blanks will appear. What is worse, since all the signals are separated, the noise is no longer attenuated by manipulating averages. This is illustrated on the map of Figure 16, where N stands for noise. A NASA colleague and I repeated this experiment, even trying (illogical though it may seem), the 1:250000 scale for various orbits and regions. The map presented here is probably the 'best' obtained, since one can 'guess' that it is of the Nile valley immediately upstream of Aswan.

(b) *Signal average.* If we program for eight signals per degree of longitude, i.e. for the 1:1000000 scale, we attenuate the influence of electronic noise, since the radiometer has distinguished at least 13 to 14 shots per degree of longitude (in actual fact at least twice this number), and the numbers printed beside each of the eight crosses represents the averages of at least two individual signals. For four signals, the importance of the noise is totally negligible. Experience has shown that, in the 1:1000000 scale, the number of individual signals per numbered point was in most cases from three to five often more, rarely less, except on the scanned margins.

This is why I made a final choice of this 1:1000000 scale. It is the extreme limit – almost mental gymnastics, as I have said. Indeed, the center regions of the areas scanned appear normal although, sometimes, empty spaces between shots left only one signal per numbered point. Towards the edges, however, the population map almost always shows a considerable drop in the number of signals; thus, noise is proportionately more important. Therefore, extreme caution must be exercised when interpreting.

Oceanographers deal in smaller scales (1:2000000, 1:5000000) which smooth out the noise problem and are satisfactory for their needs. What *we* are looking for – details (if the word can be applied to linear dimensions no smaller than 8 km) disappear on these scales. In any event, this was the largest scale I used and on this all the maps but one in the present book were prepared, except for those maps borrowed from my meteorologist and oceanographer colleagues and one stereographic polar projection map.

Using the same reasoning for the ERTS, and 'rounding out' the ground resolution to 100 m (channel 4) we perceive that separation of all signals, noise included, would

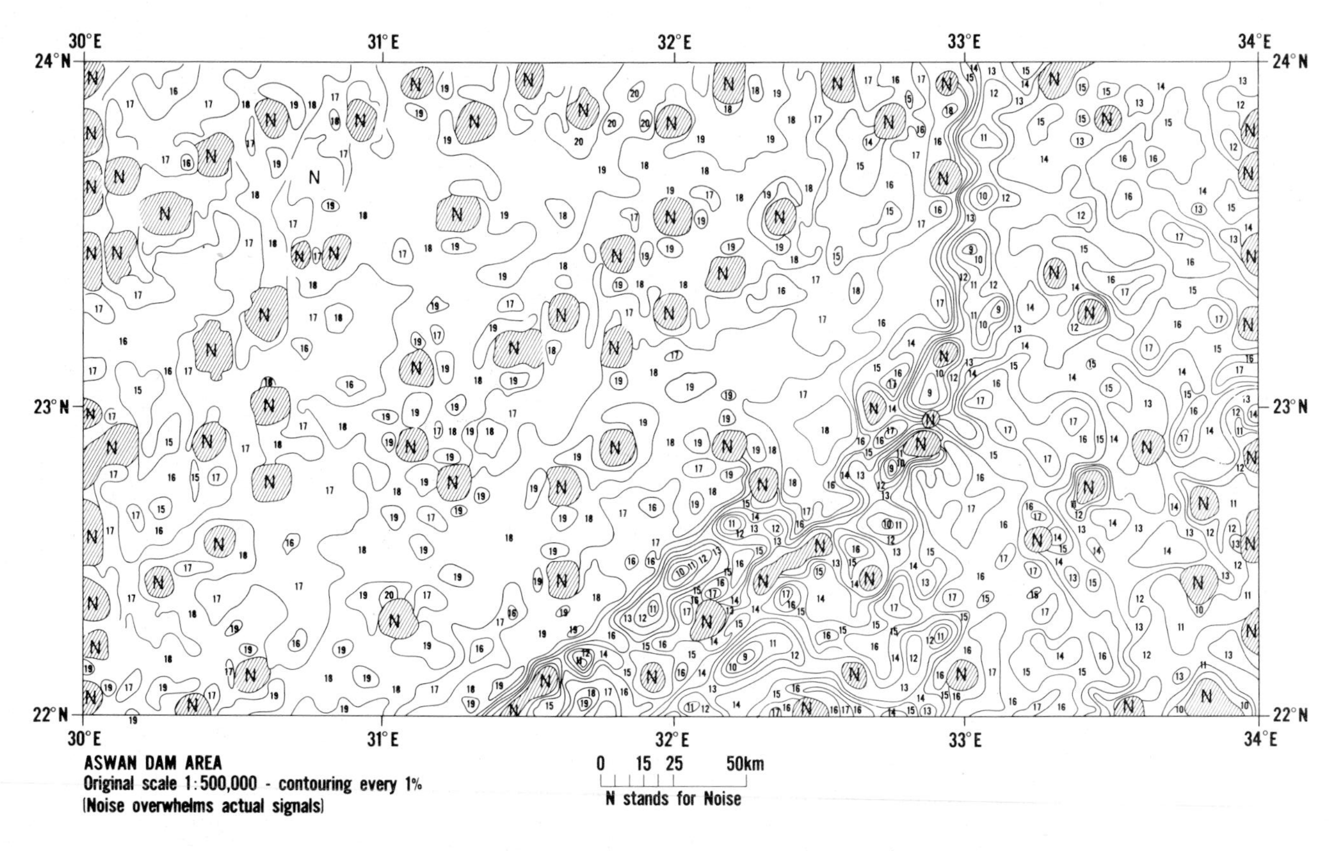

Fig. 16. Noise becomes very distinct when the scale is too large (1:500000). Infrared albedo in percent; Nimbus 3, orbit 269, 4 May, 1969. Noise is indicated by the letter N surrounded by oblique hachures.

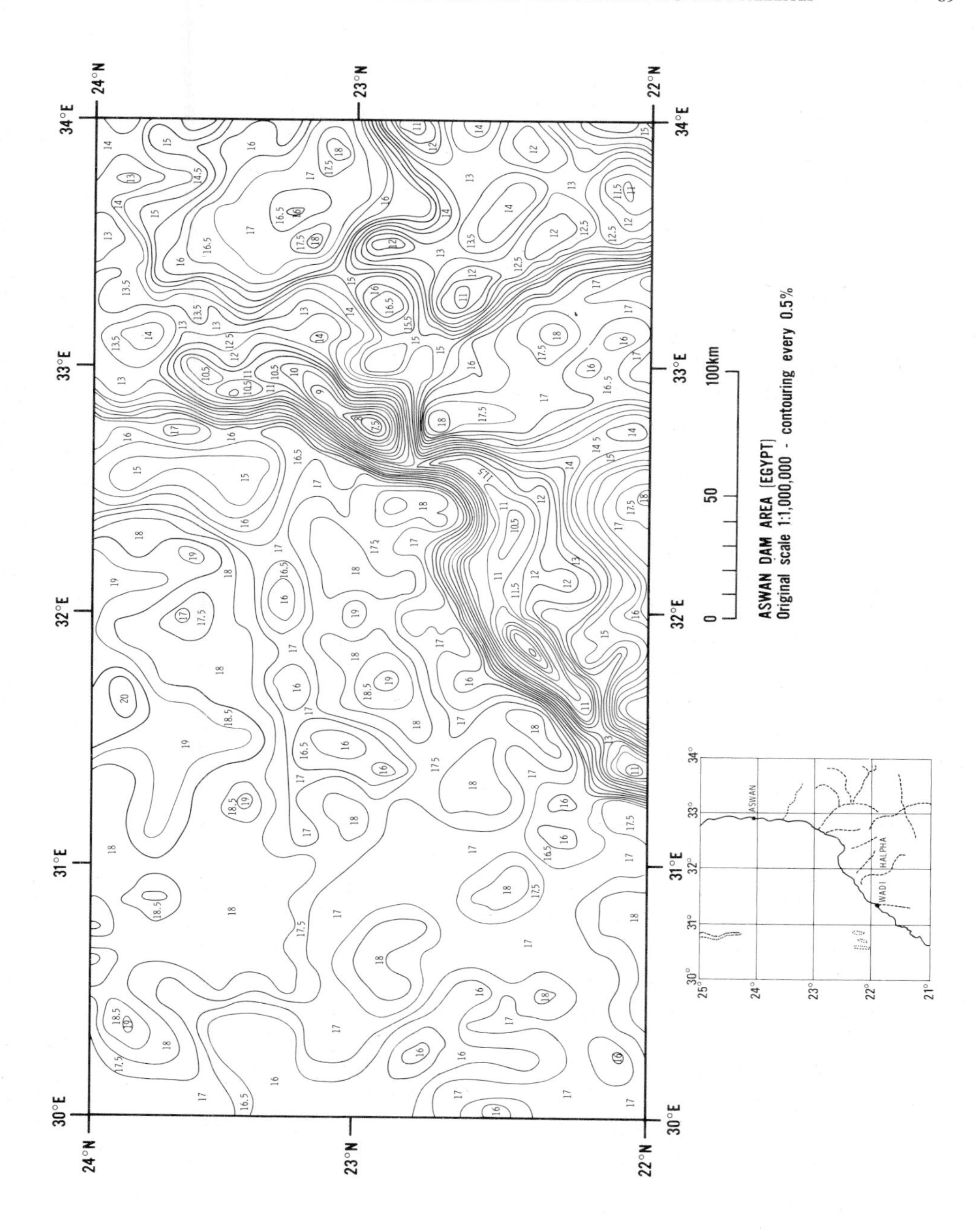

Fig. 17. Correct scale (1:1 000 000). Same orbit, same region, and same magnetic tape as for the map of Figure 16, but isolines plotted every 0.5%. – *Note:* The scales shown belong to the original computer-produced map. This note applies to all cartographic examples in this book.

authorize a scale of 100:0127 = 1:7874 (1.27 cm = one-half inch). To reduce the influence of noise, I took an average of three signals (at least) per numbered point, which comes out to a scale of 1:23662. For practical programming reasons, the final scale was about 1:25000, and I finally chose six intervals (between the crosses on the paper) per minute of longitude, 10 seconds of longitude per interval. For ERTS B, channel 5, in 1975, ground resolution being 200 m, I proposed a routine of 3 intervals per minute of longitude, 20″ of longitude per interval, or an approximate scale of 1:50000.

I write *approximate scale*, and this applies to all maps prepared on standard paper using English measurements. I hope that, due to the necessary *internationalization* of Earth Resources satellites, widespread use of the metric system will permit, among other things, computer production of cartograms on a more precise scale. Finally, let us remember that the largest scales are about 1:1000000 for Nimbus 1 to 4, about 1:100000, for Nimbus 5 (1972, SMCR), about 1:25000 for ERTS channel 4, and about 1:50000 for ERTS channel 5.

(c) *Choice of equidistance* (lines of equal temperature or equal reflectance). This problem involves choosing the proper interval between levels, and we know that this number must be greater than that characterizing photographic reproductions (photofacsimile). The thermal data are supplied at one degree interval, and reflectance data at every tenth of one percent. Thus, it is theoretically possible to plot isotherms from degree to degree and equal reflectance lines from 0.1% to 0.1%. These two extremes cannot be achieved as the final result is unfortunately the manifestation of instrumental noise. Experience, already lengthy, has proved that 10 intervals between levels gave no more information than photofacsimile. The same experience has proved that 50 intervals were a luxury not inaccessible but quite useless (still the same noise irritation) despite some very superficial successes. We finally came out with about 25 intervals, giving consistent and interesting results: this means that the isotherms are plotted every 2 degrees and the equal reflectance lines distinguish the level of infrared albedo every 1%.

I would like to conlude this chapter with a little exercise in virtuosity. I have taken the same magnetic tape (NMRT) used for the map in Figure 16, programming for exactly the same geographic limits but on the reasonable scale of 1:1000000 (instead of 1:500000) and (unreasonable) equidistance of 0.5% (instead of 1%). The result is extremely pleasing as shown in Figure 17 where all details show up sharply and noise is relegated to the background. And yet, this is none other than an exercise in virtuosity, demanding labor out of all proportion to the results obtained.

CONCLUSIONS TO PART 2

AIRCRAFT AND SPACE VESSELS: COMPETITION OR COOPERATION?

The map on Figure 17 provides me with the ideal conclusion for Part 2. I will not hide behind false modesty, and honestly say that it attests to a virtuosity acquired after hundreds and hundreds of similar exercises (several of my colleagues gave up on such attempts). But, as with many exercises of this nature, it has value as far as technical skill goes, but none whatever for results. Indeed, I went beyond the capability of the radiometer and the computer, which merely sketched in the placement of the isolines. Days on end were devoted to this labor, while a few hours would have been sufficient had I been working from aerial photographs.

And this is the whole point: what can we expect from satellites, and what part do aircraft have to play? This leads me back to the concept of ground resolution which is obviously finer from an aircraft than from a satellite. I would thus like to examine, as soberly as possible, the relative superiority of these two carrying platforms, and answer the question in the heading.

1. Satellite Superiority

Only the satellite can provide, at no additional charge, continual, round-the-clock, surveillance month after month. Fortunately, investigators are increasingly turning away from examination of one stage, one situation, and looking at developmental patterns. We have to know what is happening to our forests and crops, follow floods, observe water pollution, detect signs of imminent volcanoes in time. Tasks such as these are probably the major function of satellites – without forgetting that, due to their field of view, many (mainly geological) static phenomena are revealed which pass unobserved on aerial photographs. In the course of Part 3 we will have the opportunity to substantiate this observation.

Having said this, do we need ground resolution permitting production of maps on the 1:1000 or 1:5000 scales? I do not think so. In my opinion, and perhaps here I stand alone, ground resolution of 100 m (ERTS 1, B) is too large: the 600 m of Nimbus 5 and F seems quite sufficient as from these we can make maps to the scale of 1:100000 excellent for following the evolution of terrestrial features, excellent for showing up unexpected accidents such as smoldering forest fires invisible at ground level, volcanic eruptions, floods likely to become catastrophic, etc.

2. Aircraft Superiority

If aircraft are carrying the same instruments as the satellites, then their superiority seems to go without saying, for common-sense reasons of which I will cite only a few.

(1) The satellite is injected into orbit – and that's it. An aircraft can recommence its mission at will, as often as necessary.

(2) Cloud ceilings make the satellite inoperative with the exception of the microwaves. In the majority of cases, the aircraft can fly below the clouds.

These two reasons by themselves demonstrate the superiority of the airplane. However, I think this fact clinches the matter: only the aircraft and helicopter give acceptable resolutions for detail studies, i.e. scales larger than 1:50000.

NASA understood this evident superiority. Starting in the fall of 1970, an operational aircraft service has been provided*. This once-per-month service, equipped with exactly the same TV cameras and spectrometers as those on the ERTS, regularly overflies selected areas of the United States. I hope that all major countries will follow this example without delay.

3. Response to the Question at Issue

The answer is obvious. I will confine myself to a quotation from my own work:

> The future belongs to intimate, friendly cooperation between satellite and aircraft, the satellite doing the 'roughing out' and the aircraft the 'final polishing,' completed... by indispensable ground analysis in company with laboratory studies (Pouquet, 1972, Montreal).

* A Convair 990 and the notorious U2's (supersonic 'spy' planes).

PART 3

APPLICATIONS TO EARTH SCIENCES

With the exceptional successes of Nimbus 3 in 1969, Nimbus 4 in 1970 and Nimbus 5 in 1972, the period of 'groping' can be considered practically over: I believe that we have explored most of the opportunities offered by satellites to Earth science research.

At the beginning, little was done besides recognizing the possibilities, and only a small band of us believed in them. In 1971, the impact of the results obtained was such that large land use organizations, universities, and geological and agricultural departments have been incorporating these techniques into their research and teaching programs.

Even 1970–71 was not the breakthrough period; it was not until 1972 that the Earth sciences had satellites available. The trial-and-error period also included photographic experiments performed by astronauts onboard manned satellites. It is thus useful, before tackling the practical results obtained from Nimbus, to devote a few lines to the remarkably instructive missions of Gemini and Apollo, without forgetting that the experiments were confined to the photographable part of the electromagnetic spectrum, and without forgetting that these experiments enabled us to choose the spectral bands in which the ERTS 1 and ERTS B instruments would operate.

CHAPTER I

MANNED SATELLITES

Although the photographs taken by the Gemini and Apollo astronauts are known the world over, it will be useful here to recall some of their experimental results.

NASA organized these photographic missions with an eye to terrestrial resources. Apollo 9, for instance, was the only satellite that carried out a systematic multi-spectral photography experiment. The spaceship carried an assembly of four cameras attached to one of the portholes. These were Hasselblads 500 EL with Zeiss *f*/2.8 objectives, having a focal length of 80 mm. The shutters were manually operated. The four cameras were loaded with:

- infrared Ektachrome film, filter No. 15 (orange), NASA code A;
- Panatomic X film with filter No. 58 (green), code B;
- infrared black and white film, filter No. 89B (dark red), code C;
- Panatomic X film with filter No. 25A (red), code D.

The exposure times in seconds and diaphragms were as follows: (sunny weather):

A (orange filter): speed 1/125, diaphragm at *f*1/11;
B (green filter): speed 1/125, diaphragm at *f*1/4;
C (filter 89B): speed 1/125, diaphragm at *f*1/22;
D (red filter): speed 1/125, diaphragm at *f*1/4 or 1/5.6.

It will be noted that overexposure was carefully avoided.

The regions selected before the vessel was injected into orbit were chosen for their geological features: southwestern U.S., northwestern Mexico, southern Mexico, and Carribbean regions. In addition to these regions, the preceding flights had photographed other parts of the globe, always situated at similar latitudes.

1. Multispectral Photography (Apollo 9)

The work of interpretation consisted of comparing the four spectral bands, individually, two by two (in varying combinations), and making up a composite color by combining the three black and white plates, using colored filters (green, red, and blue). Until that time, the study had been almost exclusively visual. However, for southern California, due to tectonic features, a binocular microscope with 12 × magnification was used. It would be desirable in the future to use quantitative methods, at least for the regions of greatest interest.

The results obtained (Lowman, 1969) proved the value of multispectral photography, but also showed up its inadequacies. A few examples will suffice: one from the Imperial Valley, California; another on a region with no plant cover (Luna County, New Mexico); and another (which I will mention merely in passing), relative to a part of the Appalachians, characterized by a thick mantle of forest (eastern Alabama).

1.1. Imperial valley and environment (Plates AS 9-23-3748)

Plate A (Ektachrome infrared, orange filter). The topographic details are not well contrasted since the plant cover is scanty. On the other hand, patches of land under cultivation contrast sharply with neighboring areas and with each other.

Plate B (Panatomic X, green filter). Many details appear at the foot of the eastern mountains. The dunes in the Yuma region are clear, but less so than on plate D. The rendering of crops is only mediocre.

Plate C (black and white infrared, very dark red filter). Many erosion gullies can be seen as well as very wet soils (rivers are dark black); crops show up in detail but less clearly than on plate A.

Plate D (Panatomic X, red filter). The cultivated areas are uniformly monotonous. On the other hand, topographic features show up strikingly.

A previous study conducted using Apollo 7 photographs had led to recognition of numerous unmapped tectonic details. The Apollo 9 multispectral photograph confirmed the facts observed, which were checked in the field during the summer of 1969. The plates will be found under the number AS 9-26-3734, 3798, A, B, C, and D. Transverse faults affecting the big Elsinore feature (west of the Imperial Valley) were definitively recognized; moreover, a negative conclusion was reached as to the previously held theory of lateral displacement.

1.2. Luna county, new mexico (Plates AS 9-26-3757)

Plate A. Geological aspects, though perceptible, are less clear than on plates B and D.

Plate B. Clear geological contrasts.

Plate C. Same remark as for A.

Plate D. Striking contrasts, enabling geological aspects to be easily studied.

Only photograph D is of interest in this set. If we were to examine other desert regions, e.g. the Desierto del Altar, Mexico, photographs AS 9-26-3783, 3784, we would come out with the same results: the same is true for the Playa Wilcox, Arizona regions (AS 9-26-3753) and the New Mexico lava fields (AS 9-26-3738).

On the other hand, in eastern Alabama (AS 9-26-3790) (see Plate B, opposite page 61), the conclusions are exactly the reverse: here infrared emulsions, especially color, are superior to the black-and-white films. In eastern Alabama, the area is densely wooded and under intense cultivation. Vegetation shows up sharply on the infrared plates, as do hydrographic features. The same conclusions apply to the Mississippi Basin near Monroe, Louisiana, AS 9-26-3740, 3741, 3742, etc.

To sum up (and this is the opinion of most workers) it is always an advantage to make up a single composite color using three black and white plates (see Figure 4). In denuded terrain, panchromatic emulsions are superior, but in cultivated regions, infrared has pride of place.

From an aircraft or on the ground we can choose the best film-filter combinations, but in a manned or unmanned satellite it is essential to operate in the four spectral bands (take another look at Table IV, page 16).

2. Some Mapping Results Obtained with Photographs from the Various Gemini Satellites

Studying photographs from various Gemini satellites, Dalrymple (1970) drew up an impressive list of cartographic data. We will retain his logical classification, giving some concrete examples and the references necessary for obtaining the plates.

2.1. Geographic alterations subsequent to map publication

(a) *Lake Chad.* The maps give it a larger size than it really has. Non-existent swamps have also been mapped (Gemini 9, S-66-38444).

(b) *Timbuctu Region.* The lakes near Timbuctu have a different size on the photographs. This is probably due to seasonal flooding (Gemini 6 – S-65-63246).

(c) *Lago de Poopo* (Bolivia, latitude 19° S). Regions mapped as swampy are actually a lake, west of Machacamarca. The boundaries of Lago de Poopo are incorrect and the map shows non-existent swamps (Gemini 5 – S-65-45613).

(d) *Mouths of the Amazon.* The satellite passed over this region only nine months after aerial mapping missions; thus, impressive alterations can be seen for river sedimentation (Gemini 7 – S-65-44001).

(e) *Region South-East of Honduras and Nicaragua.* We can easily pick out at least three mapping errors, or geographical alterations since the maps were published: (1) along the Coco river where a series of meanders had not been mapped; (2) the Caratasca Lake which must be re-surveyed and mapped – the 12 (or more?) islands on the map do not exist; and (3) in 1937, meanders were mapped whose shape and position are no longer the same in 1967 (Gemini 5 – S-65-45705).

2.2. Topographical errors

(a) *Mauritania, 'Richat Structure' Region.* The rocky plateaus south-west of the Richat structure are poorly mapped, especially at longitude 13° W and latitude 20° N (Gemini 12 – S-66-63472).

(b) *River Ucayali (Peru).* The whole area (valley width, shape of meanders) was incorrectly mapped. The very placement of the river, between Glas Chaguanya and

Tarrbo de Caribos is incorrect. True, in this case the riverbed might have shifted position between 1947 (map) and 1966 (satellite) (Gemini 9 – S-66-3802).

(c) *South-East of Luzon (Phillipines)*. Lakes and channels are incorrectly represented. A peninsula between Bitaogan and Colongescon does not appear on the map (Gemini 5 – S-65-45733).

2.3. CARTOGRAPHIC OMISSIONS

(a) *Northern Mexico*. The geologic map of North America published in 1965 by the United States Geological Survey leaves out an entire volcanic area of some 520 km^2 with over 30 craters, fissures, and cones. From the ground, it is true, these forms can scarcely be seen as they are not sharply accentuated (northern Chihuahua immediately south of the border with New Mexico) (Gemini 4 – S-65-34688).

(b) *Lake Titicaca, Bolivia*. A volcanic massif extending over about 260 km^2 south of the lake does not appear on the map (Gemini 5 – S-65-45793).

(c) *Comoro Islands*. Mayotte Island is mapped without showing the coral ring encircling it (Gemini 6 – S-65-63227).

The comparison maps used for this study are on the scale of 1:1000000 (AMS [Army Map Service], Series 1301).

3. Conclusion

The above is merely a small sampling of the studies actually done, and an even tinier part of what could be done. Unluckily, these rather spectacular photographs have been used mainly for advertising or to illustrate wide-circulation publications. Earth scientists who have analyzed them logically are but few and far between.

The sole target of these photographic missions was to determine the paths to be used by satellites reserved for Earth sciences. The missions was superbly accomplished, but we cannot but note that aircraft could have produced the same results with far greater detail and accuracy.

In any event, it has been proved that multispectral photography could be used from a satellite. The fact remains, that the spectral bands inaccessible to photographic emulsions or TV cameras are of the greatest interest for remote detection of earth science features. This is something we all know, and it is thus necessary to devote the last three chapters of this text to a few results obtained from unmanned satellites in the far, the middle infrared, and also a sky approach to the microwaves recorded from a plane.

CHAPTER II

BROAD PRINCIPLES OF INTERPRETATION

Infrared has no penetrating power*. The radiations picked up by the instruments are those from the very top layer of objects, and can be reflected or emitted. Thus, the first aspect to consider is 'the surface.' Next we shall deal with the basic principles leading to interpretation of reflected radiations, the main part of the chapter being devoted to emitted daytime and nighttime radiation.

1. 'The Surface'

The ideas we are about to touch on apply mainly to thermal infrared. We measure the energy emitted by the 'surface' of objects, and we know that radiance temperatures follow Planck's law, as they are functions of real temperature and wavelength. In theory, the real temperature very strongly influences the intensity of radiated energy. In practice, many phenomena perturb the surface-atmosphere interface. A surface is a "... two-dimensional entity and, as such, its characteristics are almost impossible to measure" (Buettner, 1970).

Espousing the ideas of my colleague Buettner, the term *subsurface* is more adequate, since it designates the solid or liquid layer in contact with the atmosphere, an interface layer whose thickness is on the order of one millimeter. Radiometers and spectrometers detect the results of phenomena occurring within the subsurface and its contact with the atmosphere. These are chiefly heat exchanges – i.e. molecular, not, as within a liquid or solid mass, the result of convection currents.

The true temperature of this subsurface *could* be derived from the radiance temperature using *Buettner's coefficient* (K): $K=\sqrt{k\varrho c}$ where k represents thermal conductivity, ϱ the density, and c the specific heat of the object in question. I said 'could' quite deliberately since, as Buettner specifies, we are assuming that climatic phenomena remain constant. In practice, this is far from being the case: K changes continuously depending on meteorological fluctuations such as rain, fog, dew, moisture, variations in relative humidity, and advection currents along the interface. Under the best possible conditions we can find out k, ϱ and c easily in the case of volumes of water; but over land, these factors are more complex; they are affected by mechanical and chemical alterations, pedological processes, etc.

It is rare to be able to detect precisely the 'condition' of a given object since its

* Penetration is roughly on the order of the wavelength. With microwaves, penetration increases (on the order of a decimeter or even a meter depending on the frequency selected) which is of interest when microwaves are used for remote detection.

contact with the atmosphere is continually perturbed by the factors we listed above, or by the presence of a momentary cloud of dust or moisture undiscernable to the naked eye. All of these imponderables are constantly and irregularly altering the level of energy received by the measuring instruments. I think it well to warn users of remote detection techniques that raadiance temperature and albedo distribution on the subsurface can vary from one moment to the next. *It is a fruitless task to try to obtain "true temperatures" of objects examined. Even for ocean masses it is practically impossible and, in any case, totally useless.*

2. Reflected Infrared

For infrared accessible to photographic emulsion or for that part detectable only by radiometers or spectrometers, the basic principles are the same, and depend mainly on empirical observations.

We have already examined reflectivity phenomena and know that the albedo is clearly greater in the near infrared, from 0.8 to 1.3 μm, than in the visible, from 0.4 to 0.7 μm.

2.1. 'Photographable' near infrared

In the simple example of inrared Ektachrome film, we know that we have to deal with a color shift, as follows:

infrared is rendered by red on the positive;
red is rendered by orange on the positive;
orange is rendered by yellow on the positive;
yellow is rendered by green on the positive;
and green is rendered by blue on the positive.

Since the filters stop shortwave radiation (blue and violet), these spectral bands look dark on the positive, as the negative has not been exposed.

These color relationships are valid only if the incident energy is reflected equally in the various spectral bands. This is not in fact the case, except for inert objects with no difference between either one but the exterior coloring. Here we get two 'perturbing' factors, often related to each other: chlorophyll and water.

(a) *Chlorophyll*, which absorbs mainly at 0.45 and 0.65 μm in the visible, becomes reflective in the near infrared: this, consequently, affects the film and the *resulting red* shows that the plants' chlorophyll functions are intact and that they are in an excellent state of health. Deterioration of these functions would logically be translated by greater or lesser absorption of the near infrared, giving a weaker red shade and none at all when the chlorophyll is completely destroyed: in this case, reflectivity of the other spectral bands is no longer dominated by that of infrared and, as with all inert objects, reddish foliage shows up orange, yellowish shows up green, etc.

(b) *Water (or moisture)*. Here, the case is diametrically opposite. Water absorbs infrared except in certain cases, particularly that of water polluted often by micro-organisms which have a greater or lesser near infrared reflection capability (they show up as reddish tones on the positive photograph). In the case of moisture, the shades on the plate are close to those of the theoretical color shift given above, with one important exception: as short wavelengths are practically cut out by the filter, greenish or blueish waters appear as dark blue and black respectively.

From these basic principles, we can see the value of color infrared for studying vegetation. We can also understand that, in the domain of geological sciences, 'ordinary' emulsions are preferable except when we are trying to determine the state of water saturation of rocks or soils, or when hydrographic studies are the principal research target.

2.2. 'Non-photographable' near infrared

Here we are entering a domain that has been only tentatively explored, by Nimbus 3. As a result, we lack technical results, although the theoretical bases have been worked out in laboratory experiments and a few field observations.

The facts seem to confirm the glimpses given by these prior studies (see Part 1, Chapters I and II). For example, hot water has slightly less reflectivity than cold water, which confirms the analyses of the Gulf Stream and the Red Sea. Similarly, a study of the eastern Mediterranean showed that water turbidity off the Nile Delta diminishes the absorption power of the sea, which shows up by higher reflectance values (Pouquet, 1970, Mexico).

Outside the contrast between water/moisture (very low albedo) and rocks/vegetation (higher albedo), we are lacking in theoretical bases from which to interpret satellite observations (insufficient ground resolution). However, we may consider that dark-colored rocks such as basalts have a lower infrared albedo than light-colored rock material such as sand. Moreover, coarse-grained materials are 'darker' (greater absorption giving lower albedo) than fine-grained materials, which are 'brighter.' For example, quartz sand with an average grain diameter of 1 mm has slightly less reflectivity than similar sands averaging 0.1 mm. Finally, and here we return to our starting point, a moist soil or rock is 'darker' than the same soil or rock that has lost its original water content.

3. Emitted Infrared

3.1. Reminder of the importance of emissivity

"*After the corrections made necessary by atmospheric interference, the signals emitted in the 3.8 and 10–11 μ bands depend on the surface temperature and the emissivity*" wrote Buettner (1970, italics mine), stressing the irritating problems posed by ε: "One surface characterized by $\varepsilon = 0.9$ at wavelength 10 μm shows a temperature deviation of 7 K; another, with $\varepsilon = 0.6$ at 3.8 μm is characterized by a deviation of 12 K. These cases are extreme, but *the value of ε is vital*..." (Buettner, 1970, p. 7, italics mine).

I believe that Buettner is one of the first investigators, at least in the United States, to have called attention to this problem too often forgotten by most users (see Figure 34 page 153).

The upper part of Figure 18 shows that the presence of chemicals spilled in large quantities in this Quebec river by paper pulp works (who have found an excellent way of getting rid of their evil-smelling wastes!) is translated by a radiance temperature of some $-12\,°C$. It will be readily agreed that the real temperature, around 0° to 3–4 °C, plays a negligible role. On the contrary, a veil of mist, invisible to the naked eye, could explain the phenomenon which can be much better understood as a function of emissivity. Perhaps I should remind the reader that when Quebec was having its general strikes in the same season, the radiance temperatures became more normal,

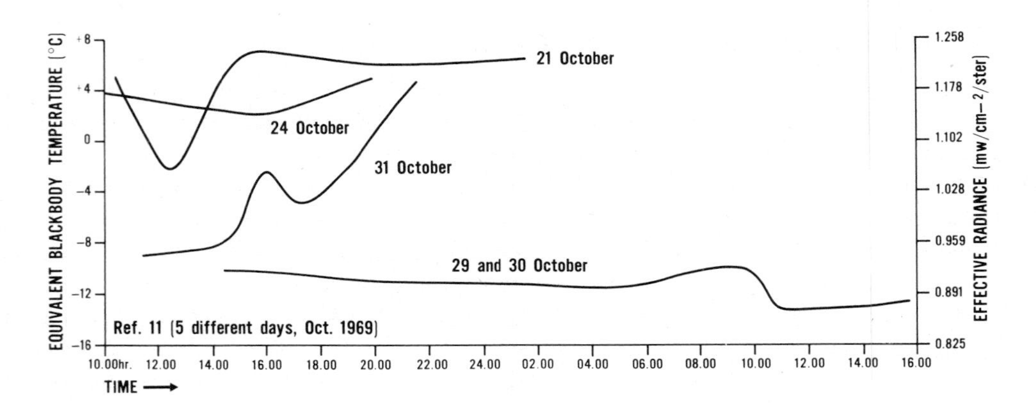

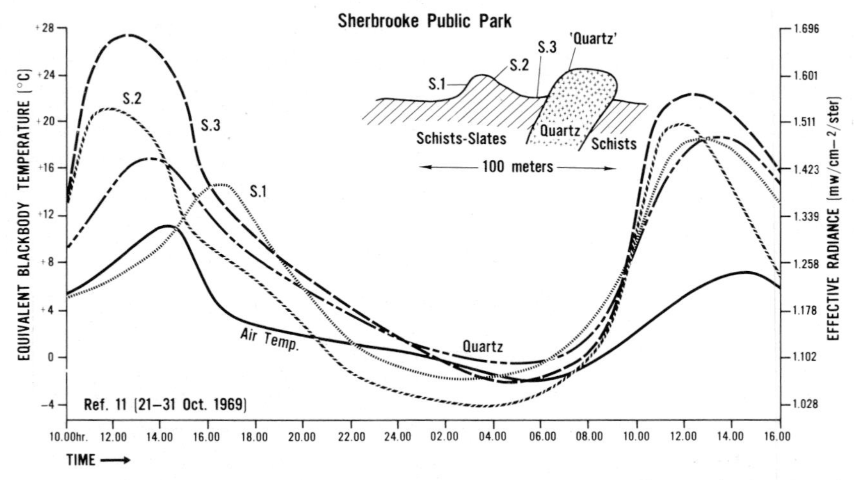

Fig. 18. Pattern of radiance temperatures measured on the ground by a radiometer in the 10.5–12.5 μm band. Radiance temperatures in degrees centigrade. Effective radiance in mw/cm^2/ster. – *Top:* pattern of water surface radiance temperatures in the middle of the River Maggog dam, Sherbrooke, Quebec, Canada. *Bottom:* comparative radiance temperature patterns of quartz and schists/slates.

close to the real remperatures, for the duration of the strikes. Such observations have been repeated time and time again, not only in Canada but also in the San Diego region where the waters of the Pacific are polluted by wastes from the United States Navy (Pouquet, *Rev. Géomor. dynam.*, 1972).

Geographers Estes and Golomb used the 8–14 μm spectral band to study the catastrophic oil slicks off Santa Barbara, California. The areas covered by oil are shown by radiance temperatures lower than those of the unpolluted water. "*...these slicks diminish air/water heat exchange and give a cold thermal picture*.... The precise nature of the *emissivity characteristics of oil on water is a fascinating field in itself*." (Estes and Golomb, 1970, p. 677, italics mine.)

It may be useful to report the results of an experiment performed by K. H. Sziekielda (at my suggestion). He added increasing quantities of sewing-machine oil to water in a tank whose volume was slightly under 0.3 m^3 and took radiance temperature readings with a radiometer operating in the 10.5–12.5 μm band, angle of view, 2°. Figure 19 shows the results: with 0.03 cm^3 we see a rise of 0.2 °C; then, with larger quantities of oil, the radiance temperature falls, rapidly at first, then more slowly, ending with a negative difference of 1.5° with 5 cm^3 of oil. At this point (out of curiosity), he added a few cc's of dish-washing detergent: the temperature shot back to within a few tenths of a degree of the starting point. We should perhaps specify that water, oil, and detergent were at the same real temperature.

We can agree, from just the facts themselves, on the *paramount importance of emissivity*. We will soon have the opportunity to see, still from the concrete facts, that, if emissivity plays the preponderant role during the day, its activity is very attenuated, relatively speaking, during the night. On the other hand, *water pollution may easily be detected* from the radiance temperatures; thus, if for this reason alone, *modern remote detection methods deserve to occupy a top position in our research techniques*.

3.2. Emitted infrared, nighttime

Remembering both the importance of emissivity and of the time when radiations are recorded on board the satellites (around midnight), the basic principles of interpretation are fairly simple (Pouquet, 1968, NASA; and 1969, Berlin).

The principal idea is that of *diurnal solar energy retention capacity*. A massive, dark-colored rock stores more energy than a sandy rock, and retains this energy for a much longer time. At midnight, the block of sandstone is warmer than its neighboring pile of sand. A coarse-grained rock is colder at midnight than a fine-grained rock of the same type.

Of all the materials, it is *water that reacts most like a blackbody*, absorbing more, retaining longer, and reemitting more. Thus, at midnight, water is warmer than any other body subject to the natural laws of heat exchange with the atmosphere.

In detail, this elementary principle of interpretation must take into account the materials' conducting properties, structure, porosity, density, etc. Also, as we have already seen, the subsurface has neither liquid nor solid characteristics and is affected by local phenomena that we have only partially enumerated. Nonetheless, other things being equal, the radiance temperature of water is always higher than that of other bodies during the night.

This leads to the *second principle* of paramount importance in the field of applied geography: *any moist body is always warmer at midnight than a similar but drier body.*

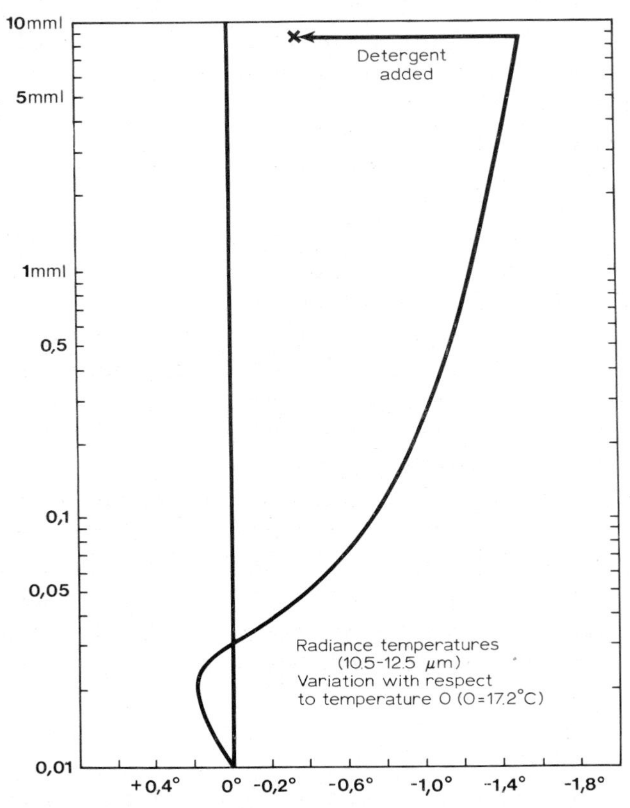

Fig. 19. Influence of oil films on the water surface. – Ordinate: in cubic centimeters, on a logarithmic scale. Abscissa: temperature deviation with respect to original radiance temperature. 0, the real temperature, was the same from the beginning to the end of the experiment. Value 0 = 17.20 °C; Free surface: about 1 m^2; depth, about 30 cm. (Graph communicated by K.-H. Sziekielda. Spectral band 10.5–12.5 μm.)

It is thus easy to detect water-rich or water-poor formations, formations more easily impregnated by moisture, and those in the paths of condensation water, floods, underground water seepage, etc. This is the second principle which leads to recognization of former valleys, provided alluviums exist.

In this connection, we must remember one all-important phenomenon: *soil moisture always moves in the direction of the colder zones in soil profiles**. In the daytime,

* On this subject, see: R. F. Miller and K. W. Ratzlaff, 'Chemistry of Soil Profiles Indicates Recurring Pattern and Modes of Moisture Migration', *Proc. Amer. Soc. Soil Science*, 1965. J. Pouquet, 'Water Movements in Soil Surface Layers', *Bull. Assoc. Géogr. Fr.*, 1965.

the Sun's heat 'pushes down' the moisture in the soils, which store a fraction of diurnal solar energy. We can thus see why moist soils cannot be detected in the daytime unless, instead of 'pedological' moisture, we are dealing with widespread irrigation. In the night, the subsurface cools off rather rapidly and the soil moisture migrates upward toward this area, thus transfering the stored energy and giving this subsurface area slightly higher radiance temperatures.

This second principle permits easy recognition of soils subject to sheet erosion. In this case, the 'intact' sectors, i.e. those that still have an A_0–A_1 horizon, are warmer than those where the A_2 horizon (one with poor moisture retention), crops out.

The *third principle* is straight forward: *an altered rock (one that has undergone chemical or mechanical erosion) is warmer than an unaltered rock*. With rare exceptions, alteration translates into darker shades and also into granule size modification, due to the relative increase in the content of fine particles favorable to moisture retention. We can see how easy it becomes to distinguish recent from old dunes, and an old eroded surface (numerous surface deposits) from a recently formed surface in which the bare rock crops out.

We are here touching on the *fourth basic principle: bare rocks are distinguished from one another* by texture, physical, and chemical properties, color, etc. In combination with the second principle, this explains why tectonic areas stand out on radiance temperature distribution maps. In theory, we should be able to trace the lines separating geological outcroppings, but the facts are otherwise.

First of all, humidity obscures this differentiation between rocks. Secondly, during the night, the *real temperature*, a function of the solar energy retention capacity, and, in addition, *height above sea level* in its turn obscures the emissivity power, something that every map prepared from satellite information shows without exception. Consequently, during the night, height differences, and thus heat differences*, play a considerable part. However, it is rare for the heat difference due to height to be shown by a radiance temperature difference, the latter being almost always 'hotter' or 'colder' than it should be because of altitude alone.

Here, we are arriving at the real key to the interpretation of nocturnal radiance temperature distribution maps: *when he finds heat anomalies, the interpreter seeks by a process of elimination to determine the cause.*

3.3 Emitted Infrared, Daytime

Field experiments using a radiometer in the 10.5–12.5 μm band (Pouquet, 1972, *Rev. Géomor., dynam.*) and the results in this spectral band from Nimbus 4, launched in April 1970 (Pouquet, 1970, Mexico) lead to an outline of the basic principles.

If we take another look at the lower half of Figure 18 above we note one of the advantages of radiation emitted around midday over that recorded around midnight. At 12:00, the contrasts are sharp; at 24:00, they flatten out. In the middle of the day, schists/slates are warmer than quartz, with the exception of one of them, S^1

* Wet adiabatic lapse rate 5 °C per km; dry adiabatic lapse rate 10 °C per km.

which, due to its topographical position, was still partially in the shade. In this example, the S^3 and quartz samples are the most significant since they have the same orientation and were seen from the same angle of view. We see that schist has a temperature of 10–11 °C (sunny weather) or 4–5 °C (overcast sky) greater than that of quartz. Here, I should point out that the satellite could not record the greatest contrast, as the cloud ceiling on the following day was an insurmountable obstacle.

Shortly after sunrise, about 7:00–8:00, quartz is warmer than schist. At midnight, the two rocks are at about the same temperature. These observations, incidentally, have an exceptional value: if, instead of two measurements in 24 h, we could have a series of observations (enabling us to draw curves similar to those in Figure 18), it would be possible to identify most rock, soil, and plant formations provided we had control curves in sufficient number; the parameters of such curves would be stored in the computer.

We also lack field observations to draw up a comparative list of diurnal radiance temperatures of multiple rock samples. I thus consider it preferable to confine myself to some elementary principles, all derived from actual observations, both in the field and from the Nimbus 4 radiometer measurements.

Principle No. 1 is based on water and moisture (but not soil moisture). In this case, moist objects are distinguished by radiance temperatures sharply lower than those of their surroundings.

The second principle is linked to emissivity. The effects of height differences give way to this factor which now occupies the top position. As a result, the distribution of geological outcroppings, or of plant formations as the case may be, is strongly accented.

For the same reasons, we arrive at *rule No. 3*, somewhat comprable to principles 3 and 4 of nocturnal emitted infrared. This third principle is difficult to express. We can state it as follows: at midday, *light objects are warmer than dark objects*, with three important exceptions. First, this remark only applies in sunny weather, the only kind of weather in which the satellite's radiometer can take measurements. Second, this rule is observed and modified by petrographic textures. A fine-textured rock is colder than the coarse-textured rock of identical composition, and vice versa. Thus, a dark, fine-textured rock can be at the same temperature level as a light, very coarse-textured rock. Finally, if the sun has been out only a few moments before the satellite passes (sudden cleaning of a cloud ceiling) we are in a situation halfway between nocturnal and diurnal radiation characteristics.

We can see that this raises a sensitive problem of interpretation. However, to confine ourselves to the practical aspects, we should readily agree that these daytime measurements are of enormous interest to the geological, pedological, hydrographic, etc. sciences. In this respect, I would place the pedological consequences in the foreground. An eroded soil is substantially warmer than the identical soil unaffected by sheet erosion. In this case, the light-colored horizons are exposed to the Sun and heat up faster than the A_0–A_1 horizon, which is usually darker and moister.

Unhappily, if these daytime measurements are valuable to the earth sciences in the restricted sense of the term, the same does not appear to be the case for oceanography, if we are to judge by the results from Nimbus 4. At midday in sunny weather, a thin veil of water vapor and mist obscures every anomaly on the ocean surface. Some exceptions confirm the rule, if indeed it is a rule, The eastern basin of the Mediterranean has shown that water turbidity caused by sediment swept out by the Nile shows up by a 4–10 K increase in radiance temperature. Unfortunately this phenomenon has been observed only on ten or so digitized maps, which is all too scanty for even an attempt at an interpretive rule.

4. Passive Microwaves

Only a limited amount of observations have been recorded in the passive hyper frequencies; however, we can already propose some principles of observation.

First principle. Based on the tremendous variations in emissivity values, linked to water, humidity, moisture, day or night, under a cloudy or clear sky. The average emissivity is around 0.4 for water, but this value increases when the water temperature decreases and vice versa. When an object is wet, its brightness temperature immediately drops very sharply. Looking at two identical soils, one dry, the other slightly wet, the latter shows a much lower T_{Br} than the first. Finally foreign bodies poured into water (dust and /or chemicals) provoke a sharp emissivity increase.

Second principle. Contacts between different rock formations are sharply emphasized, due to the emissivity differences. One sees the geological facilities helping with geological maps; one sees too, how tectonic accidents could be easily detected, the difference in emissivity between rock formations on either side of the 'lips' (fault line) being 'boosted' by water: joints and faults are sites favored by the migrant humidity.

Third principle, linked to recognition of soil formations. This principle is the same as for rock formation, but it goes farther: the moisture of different soils and that between different soil horizons shows up with much lower brightness temperatures. The passive microwaves are ideal for the remote detection of the most insidious type of soil erosion, namely sheet erosion.

Fourth principle. All man-made constructions, buildings, highways, factories, etc. considerably reduce emissivity.

Negative aspect: As far as I know, it does not seem that plant cover, crops, meadows, etc. are depicted with better accuracy than with 'classical' thermal infrared.

To sum up, passive microwaves are unrivalled for detecting water pollution, soil erosion, and geological-geomorphological features. To use one of our catchwords once again, environmental studies have a great deal to gain from the passive hyper frequencies.

Despite the small number of studies performed based on satellite records or infrequent ground observations, some major principles have come to light. These principles are still in a formative state: they must be amended and, even more important, their number must be increased by the quantitative growth of analyses of digitized maps and field experiments.

Provided we *never consider the values taken from electromagnetic signals as definitive*; provided we *do not attempt a futile extrapolation of real temperature data*; provided we *are not the slaves of ideas, laws, prejudices, theories*... the analysis of infrared reflected or emitted, and passive microwave radiations can lead to better knowledge of the phenomena affecting the surface of the globe.

CHAPTER III

GEOPEDOLOGICAL EXAMPLES*

All maps but one illustrating this chapter were computer-produced on the original approximate scale of 1:1000000 which is the greatest possible having ground resolution of only about 8 km.

As previously stated, we are not dealing with 'definitive' results enriching geopedological sciences. The examples that follow are an illustration of the present capabilities of satellites, but are not considered to be anything new. The maps and brief comments thereon, which are the subject-matter of this chapter, will show that remote detection, particularly from satellites, is no more than a new and powerful tool. which in no way claims to replace our traditional research methods – quite the contrary.

1. Reflected near Infrared (0.8–1.3 μm)

Experience has shown that the spectral region 0.8–1.3 μm offers hitherto almost unsuspected possibilities. I think the successes scored by the Nimbus 3 high resolution radiometer daytime channel are the logical outcome of the sharp contrasts always recorded in the middle of the day (intensity of reflected energy, spectral quality of radiations).

1.1. Río Paraná – Río Paraguay Region

The map of Figure 20 (Pouquet, 1970–1971, NERO) enables us to understand the geopedological possibilities of the spectral region in question. The few hypotheses that follow were worked out through analysis of a fairly large number of digitized maps based on Nimbus 2 (emitted nocturnal infrared), and Nimbus 3 (reflected diurnal near infrared) information. We can classify the main results into two categories: seasonal changes and permanent features, though the boundary between these two categories is somewhat arbitrary.

(a) *Seasonal Changes.* At the end of the southern summer, we observe a great increase in moist soil, marsh, and flooded areas; all of these features are toned down in midwinter. In the spring (the situation shown in Figure 20), the so-called 'permanent' (geopedological) features are strongly accentuated. All of the places where the vestiges of previous floods are still visible correspond to distinct pedological units. These are, for example, the lakes of the upper Río Corrientes, southwest of Encarna-

* GEOPEDOLOGY: The science which associates geology, geography, and pedology. Good alternatives to this word are: applied geography, applied geology, and active environmental science.

ción; the water basin located immediately northeast of the point 60° W–28° S. We find, in sharp relief, what I have called (NERO report) the 'island massif' (emphasized by higher reflectance characteristics) stretching between Goya and Corrientes (Paraná-Paraguay confluence); this region is always in sharp contrast, due to infrared albedo values substantially above those of the surrounding areas.

(b) *Permanent features.* These we have in plenty, but all must be considered to be supporting provisional theories that only field research can contradict, confirm, or amend.

(1) *Former extension of the Río Paraquay* from San Pedro, upstream to the foot of

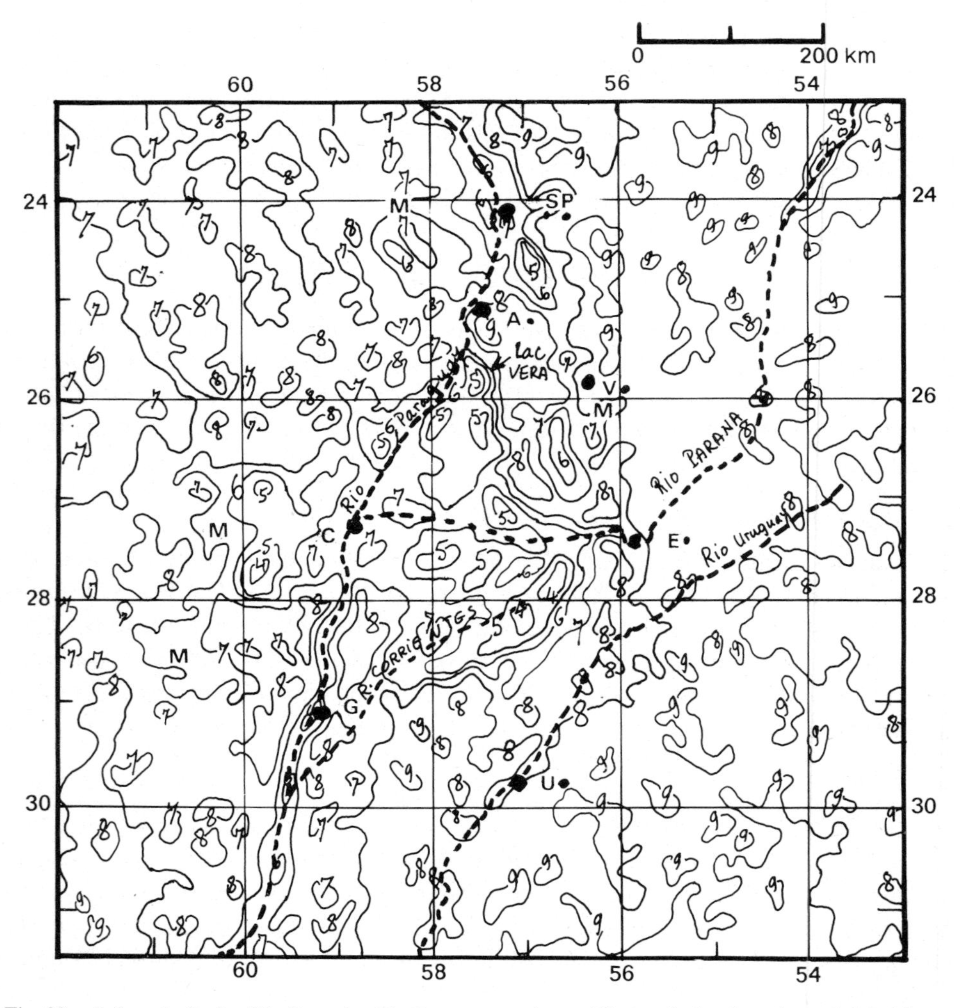

Fig. 20. Infrared albedo. Río Paraná – Río Paraguay region. – Nimbus 3, daytime channel, 0.8–1.3 μm; orbit 2618–2619, 26 October, 1969. A: Asunción; C: Corrientes; E: Encarnación; G: Goya; M: marshes; S.P.: San Pedro; U: Uruguaiana, V: Villarica.

the hills east of Asunción. This was perhaps a former lake fed by the river, poorly drained, and gradually shrinking.

(2) *Former course of the Río Paraná* between Encarnación and Goya. Precisely along the extension of the valley, upstream of Encarnación, the Río Paraná probably flowed along the valley presently occupied by the Río Corrientes.

(3) *Flood plain common to the Río Paraná and the Río Paraguay*. This flood plain is shaped like an enormous triangle with its apex towards Encarnación and base at Asunción to the Paraná-Corrientes confluence. We can see the remains of a much more lavishly-watered past; at the same time we can discern the present trend.

The entire region can be characterized by a banal expression 'uncertain drainage.' Certain aspects appear to be confirmed – in particular, the shriking of this vast flood plain. Some of its sectors, the 'island' for example, are no longer subjected to annual floods.

This brief summary stresses the capability of the operation of this radiometer in the near infrared. It shows, above all, that problems whose solution demands further aerial observations and field work can be brought to light; this is more essential than ever.

1.2. Region of the Niger Inland Delta (Macina)

Figures 21 and 22 show the situations as of July 23 and August 24, 1969 *. Here again, these two 'conditions' are part of a pattern studied using Nimbus 2 (1966) and Nimbus 3 (1969) information, throughout the lifespan of these two satellites.

First, we can observe the northbound progress of the rain front; from a latitude less than 13° N on July 23 it moves to about 14° N one month later, as seen by the northbound extension of regions watered by winter rainfall (greater absorption of near infrared by moist land in the south; stronger reflectance of still-dry surfaces in the north). The phenomenon is particularly obvious in the right-hand half of Figures 21 and 22.

More important, in my opinion, is the fact, shown up by the northwest-southeast oriented rectilinear alignment reaching the Delta halfway between Mopti and Lake Faguibine. This narrow ruler-stright strip continues the alignment of the escarpments bordering the Hodh basin but, while at the northwestern end it follows a visible geological contact (different rocks), thereafter it elongates and continues within fairly uniform formations. This is very likely a tectonic feature of enormous amplitude.

Of greater geopedological and paleogeomorphological importance are the following observations, checked time and time again in May–November 1966 and April–November 1969.

(a) *Former course of the Niger, east–west*, with probably a huge flood area, to the immediate south of the tectonic feature, former course penetrating into the Hodh basin.

* To make the maps more readable, color in surfaces with reflectance equal to or less than 8% in red; 8–12% yellow, and 12–14% blue. Thicken isolines for 9, 10, 11, and 12%.

(b) *Former course of the Niger, south-north*, leaving the present valley in the Segou-Sansanding region and flowing north. This former course may be plotted from numerous low-reflectance islets, spaced out but always apparent from one map to the next.

These two facts enable us to piece together the recent pattern of development and to make an extrapolation into the future.

The Niger, coming down from Fouta-Djalon, used to flow as far as the Segou

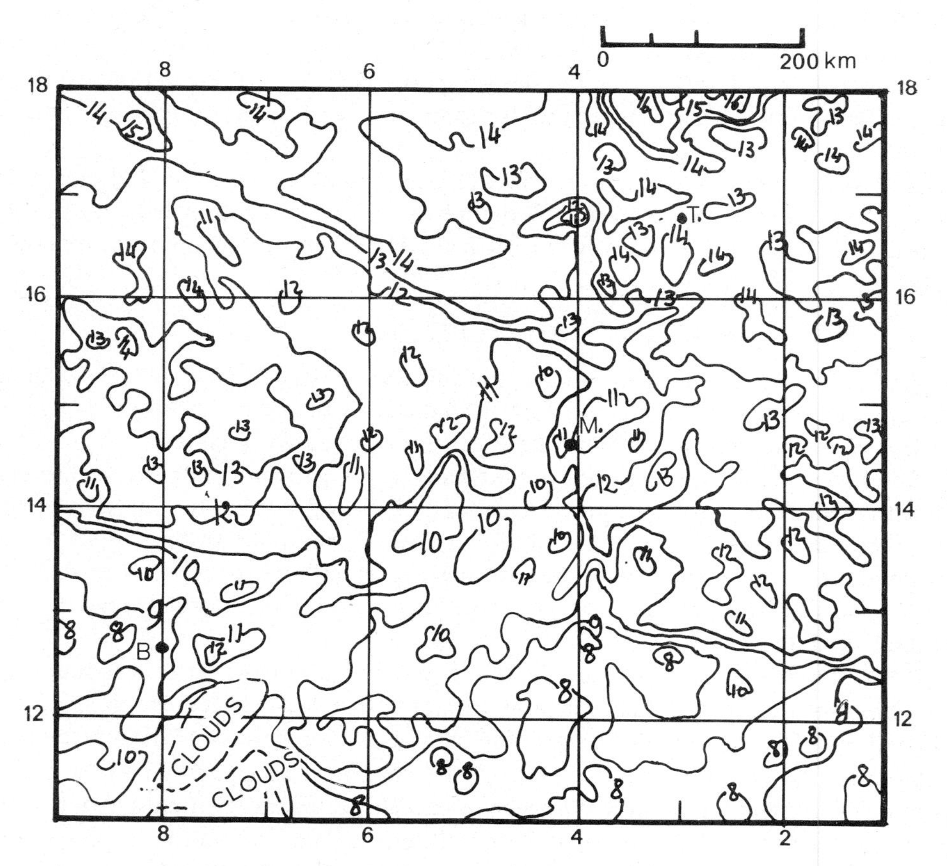

Fig. 21. Infrared albedo. Region of the Niger inland delta. – Nimbus 3, daytime channel 0.8–1.3 μm, orbits 1074 (23 July, 1969) and 1772 (24 August, 1969).

region. Here the river forked towards the north, continued in this direction for some 200–250 km, then turned northwestward and entered the Hodh basin. A climate of recurring dryness prevented the Niger from following this course. It flooded the Mopti region, thus creating the first inland delta. At the same time, due to its dried-up flow, an alluvial fan was formed sealing off the north-south route for good. A final

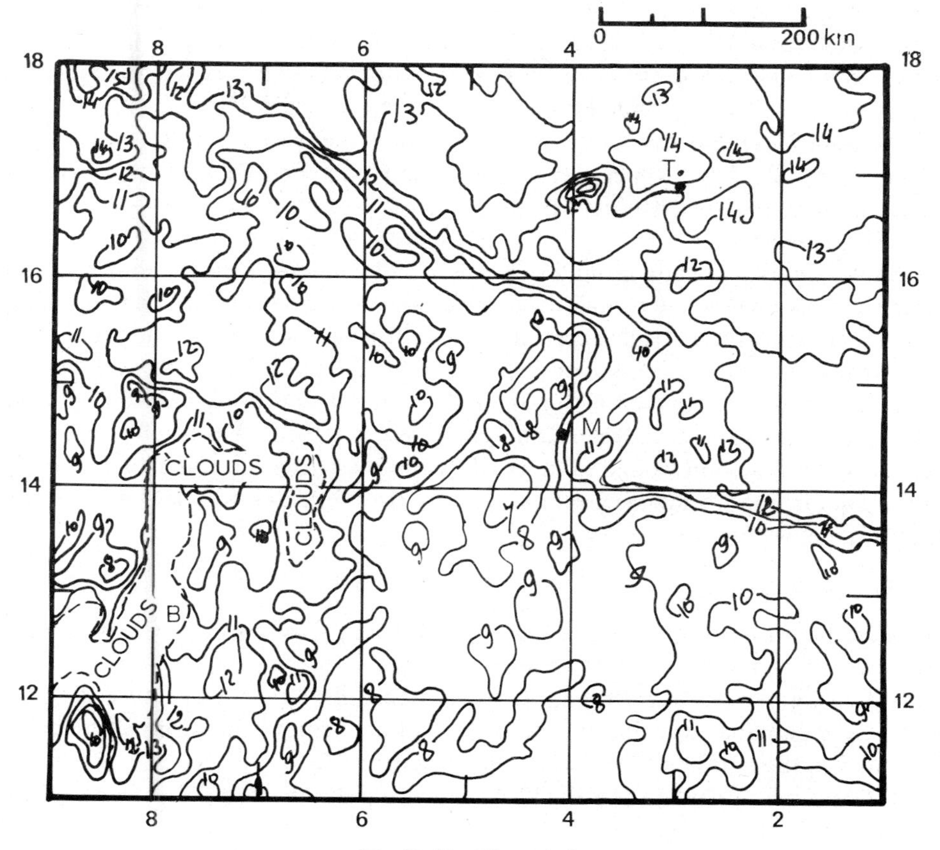

Fig. 22. (See Figure 21.)

wet climatic phase permitted the Niger to leave the Delta about 100–150 km north of Mopti, going back to its former southeast-northwest bed*.

We know that, about 3000–5000 years ago, the inland delta was finally emptied by capture to the benefit of an west-east flowing river tributary. This river tributary bears the name of the river, having created the inland delta, and, as a questionable feeder of the Senegal basin, has become the upper branch of the present Niger.

The entire Macina region, with its marshes and shifting river beds, could be proposed as a classical example of hydrographic uncertainty. A large part of the water flowing from the upper Niger is lost for the middle and lower Niger. 'Lost' must be understood as a consequence of evaporation, infiltration, and underground streams following the former riverbeds through alluvial deposits.

We know that former river channels containing alluviums, provided they are not too much altered or compacted, provide preferred drainage paths for infiltrating

* Further details may be found in my Allied Research Associates report, NERO publication (see Bibliography, under Sabatini).

waters of whatever origin. At the time of the winter seasonal high-water levels such underground passages drained off the excess water. In the case in point, this fact is shown by a tendency toward lower infrared albedo values and, during the night, by higher radiation temperatures by reason of the increasing moisture content along these ancient watercourses). In simpler terms, I think that part of the delta water finds an outlet through these former valleys leading to Hodh and perhaps beyond. Such a fact could have practical aspects that may be suspected by comparison with another region of the world.

I would suggest a careful ground survey guided by the satellite observations along these presently abandoned routes, using the classical methods of sounding, sampling, analysis, statistical treatment, experiments, etc. It is my profound feeling that all of these former valleys could be converted into what I like to call *African Imperial Valleys*. Just as the waters of Colorado and Northern California have brought unprecedented prosperity to the *California Imperial Valley*, the volume of water necessary for irrigation would be supplied by the West-African 'water-tank,' i.e. the Fouta-Djalon and neighboring mountains, via the Niger and the Bani, to cite only the two chief waterways.

1.3. Southern France

I suggest that the reader use this last example (Figure 23) as an exercise. I will confine myself to a few non-explanatory remarks.

(a) As may be seen, the Alps appear with light tones (reflectance over 10%) not along the mountain ridges but on the margins. For example, the Mont Blanc group is located in the middle of low reflectance values.

(b) The same applies to the Pyrenees which, unlike the Alps, do not recall the form of a mountain range.

(c) With a few exceptions, the Rhone Valley is surrounded by nuclei of clearly lower reflectance (less than 7–8%).

(d) The Aquitaine Basin deserves special attention, I believe, especially the Landes and the patches of different reflectance found there.

(e) Finally, when high reflectance values are fairly isolated (Saragossa region for example) and have geometrical shapes, they often consist of more or less scattered cloud patches.

2. Emitted Nocturnal Infrared

More detailed studies have already been published (Pouquet, 1967, 1968, 1969, and 1970) and I believe that the case of nocturnal infrared is now fairly well understood. I will thus confine myself to a single example, that of the Caspian-Aral region, Figure 24, noting once more that several 'orbits' covering the same region have been analyzed.

Among numerous other observations, those of the eastern regions of the Caspian Sea area seem to me to be of the greatest geopedological and paleogeomorphological interest. In this respect, two facts must be brought out.

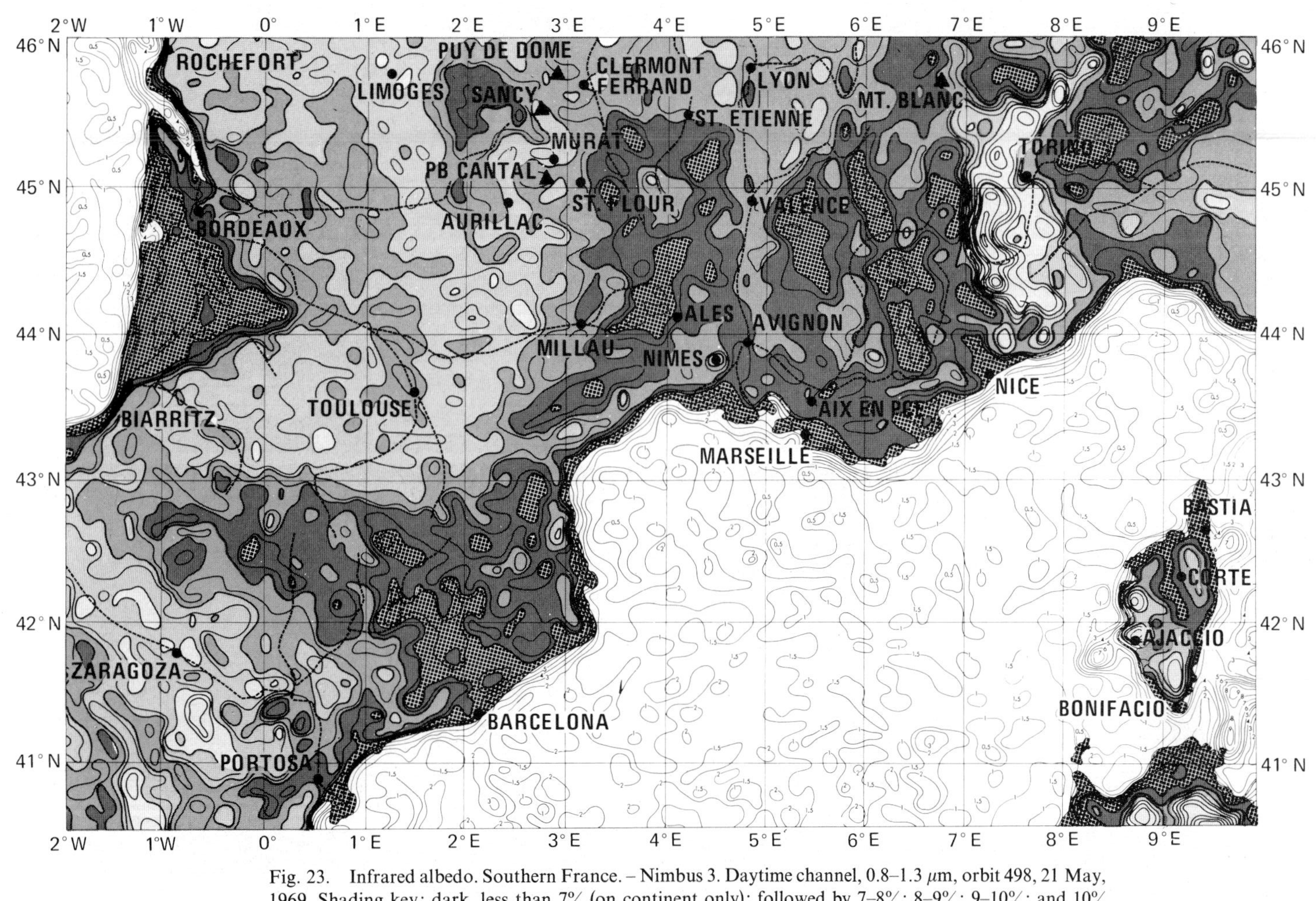

Fig. 23. Infrared albedo. Southern France. – Nimbus 3. Daytime channel, 0.8–1.3 μm, orbit 498, 21 May, 1969. Shading key: dark, less than 7% (on continent only); followed by 7–8%; 8–9%; 9–10%; and 10% and above.

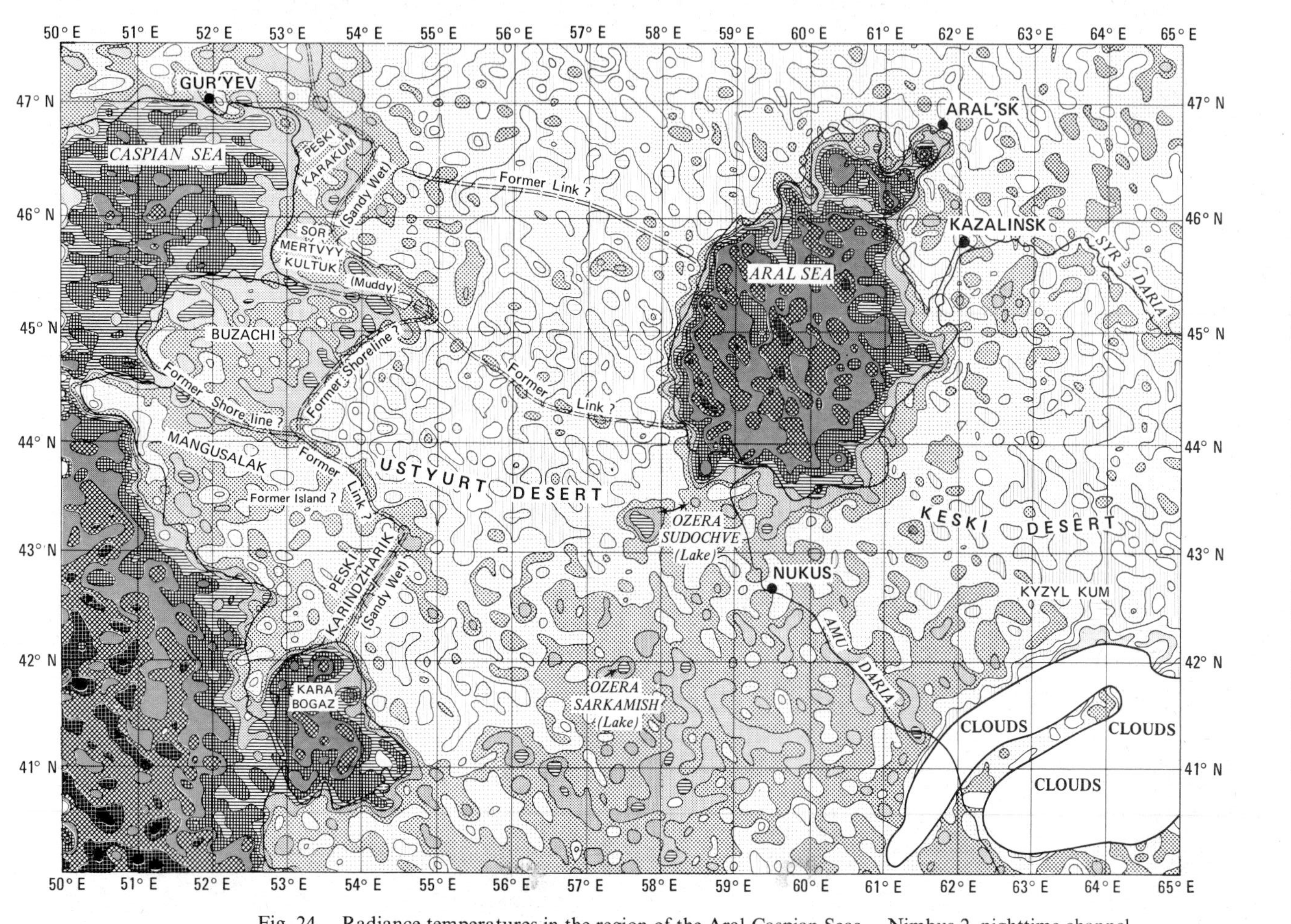

Fig. 24. Radiance temperatures in the region of the Aral-Caspian Seas. – Nimbus 2, nighttime channel, 3.5–4.2 μm; orbit 1498, 4 September, 1966. T_{BB} = equivalent blackbody temperatures (radiance temperatures).

(a) *Non-geometrical areas with higher radiance temperatures.* You can see that I am alluding to the entire region north of the Kara-Bogaz Gulf. This is mainly a marshy area with an occasional, moist clayey-sand outcropping.

(b) *Rectilinear gradients and alignments of islets with higher radiance temperatures.* These alignments delimit a huge area east of the present shores of the Caspian Sea. By way of a hypothesis (supported by several Soviet colleagues to whom I have sent many of these maps) I am tempted to see these as former shorelines, attesting previous extension of the Caspian Sea which upon receding, exposed some areas of high relief.

In addition, and this is something constantly found on most of the Nimbus orbits, one can imagine a former link between these two inland seas; perhaps even a second at a higher latitude.

For the record, I will cite one negative fact: Syr-Darya and Amu-Darya do not appear on this map whereas, with reflected infrared, these rivers are faintly (Syr-Darya) and strongly (Amu-Darya) brought out. On these same maps in the 0.8–1.3 μm spectral band, the tectonic features in the immediate vicinity of Nukus show up particularly well (Pouquet, 1970, San Francisco and Tucson). The reason I make this comment is to draw the reader's attention to an imperative rule for all remote detection studies: the need to analyze as many orbits as possible for the region in question, by analyses of the various spectral bands we have available.

3. Emitted Diurnal Infrared

Figures 25a and 25b give an excellent example of what may be expected of such measurements, taken around midday.* For once, I think it will be profitable to enter into the analytical details. Moreover, this region was covered extensively by Nimbus 2 and Nimbus 3. This map is the only one for which I can speak affirmatively, as the studies performed with the aid of these two satellites were checked in the field in 1967–1968, in 1969, and in 1970.

3.1. Some characteristics of the terrain

(a) *Influence of topography.* Most of the mountain ranges are visible, but not all. With the exception of the Sierra Nevada, the cold patches corresponding to high relief represent petrographic features rather than height differences.

The *coastal ranges* are not distinguished, despite their considerable altitude. The San Gabriel and San Bernadino Mountains, and even the San Gabriel 'Little Mountain' are immediately recognized: something that the small differences in relief cannot explain.

The San Jacinto mountains west of Palm Springs are well delineated, but not the Santa Rosa Mountain west of Mecca, while a band of low radiance temperatures trends southward parallel to the Santa Rosa mountain, but in a depressed position.

* These results, with others, were presented in November 1970 at an international conference in Mexico on arid lands. One of the two maps appears in the appendix to J. M. Ribot's thesis (1970).

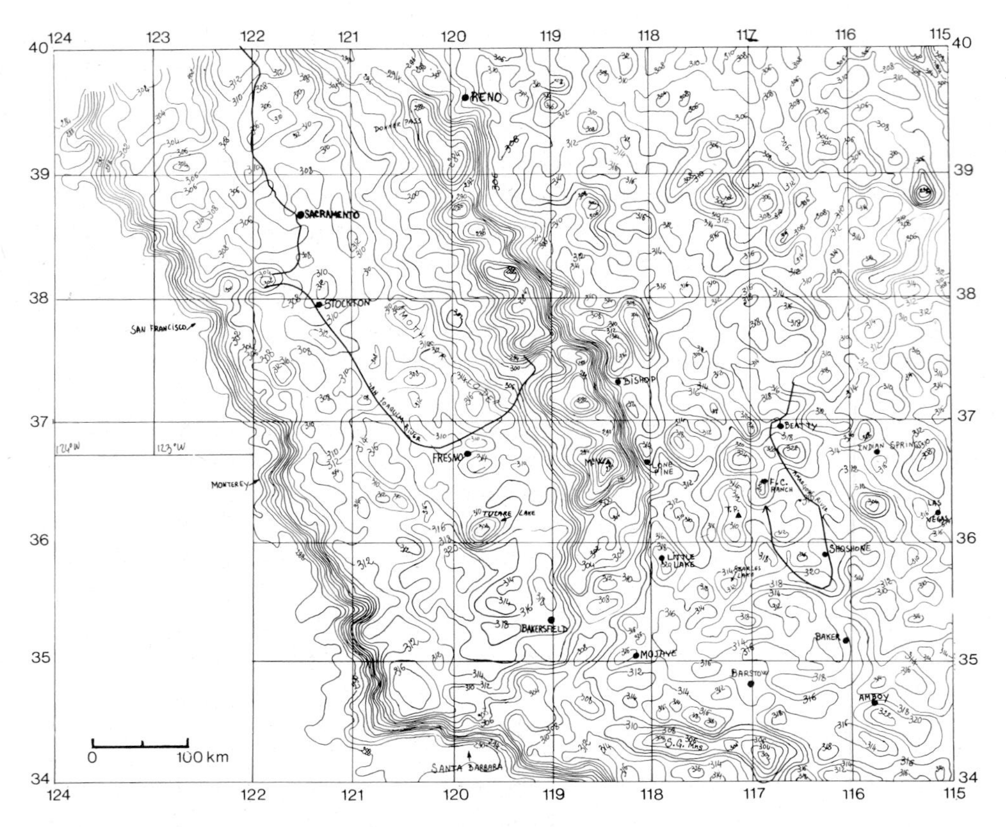

Fig. 25a. Diurnal radiance temperatures. Southwestern United States and Northwestern Mexico. Nimbus 4, channel 10.5–12.5 μm of the THIR high resolution radiometer; orbit 503, 15 May, 1970. Radiance temperatures in degrees Kelvin, isotherms at 2 degrees intervals. *F. C. Ranch:* Furnace Creek Ranch, Death Valley; *MtW.*: Mount Whitney, Sierra Nevada; *T.P.*: Telescope Peak, Panamint Mountains.

The Sierra Juarez west of Laguna Salada, is shown by a very steep temperature gradient, in total disagreement with the contour differences. The Spring Mountains west of Las Vegas are also characterized by a very steep gradient. Between the peak and Indian Springs Valley, there is a decrease in altitude of about 2500 m; assuming an adiabatic lapse rate of 0.7°/100 m, the crest line should be 17° colder than the Valley which actually has a radiance temperature no higher than 12°*. By comparison with the Sheep Range north of Las Vegas and Spring Mountains, the temperature gradient coincides with the adiabatic gradient (4°). Field work has shown that Indian Springs Valley is colder due to irrigation, while the mountain ranges, due to their petrographic similarities, fit in with the temperature differences as a function of altitude differences.

* For a height difference of 2500 m, the extreme lapse rates are: 12° in the case of wet adiabatic, which is apparently not the case; and 25° in the case of dry adiabatic, nearer to reality in this semi-arid region.

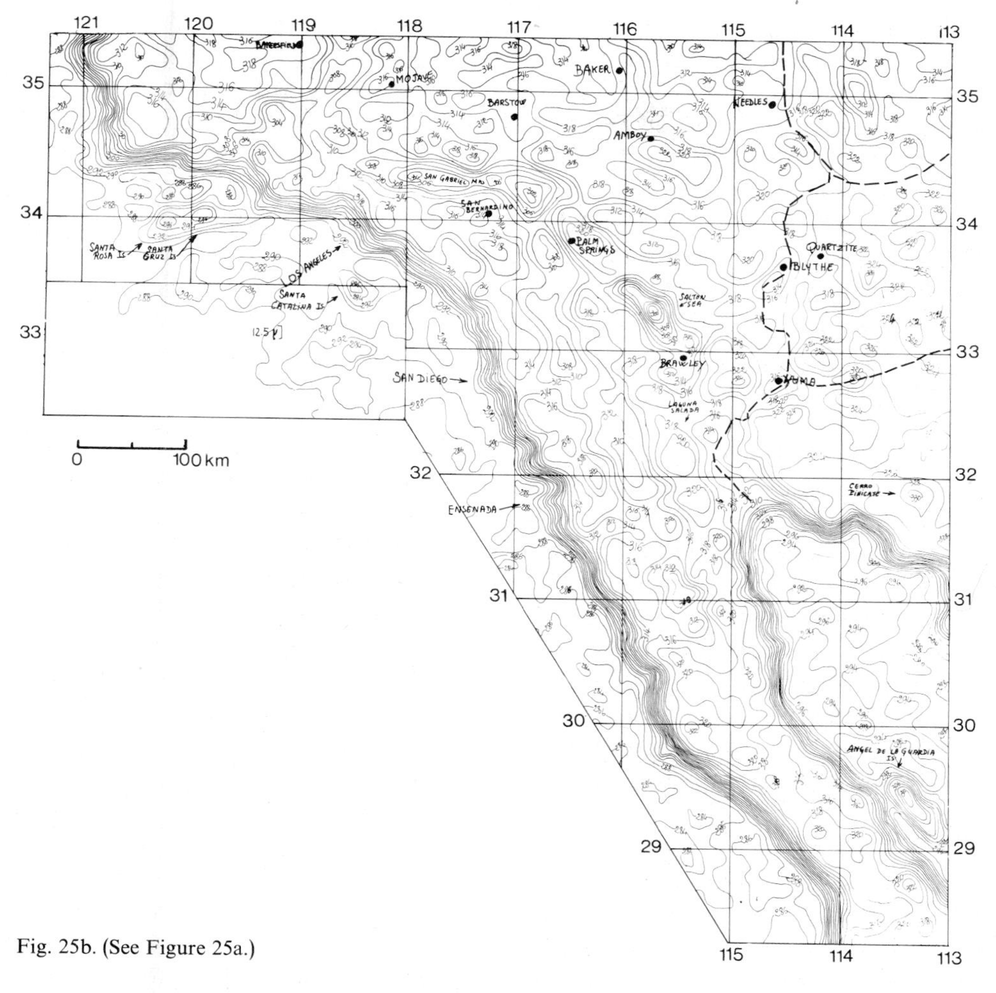

Fig. 25b. (See Figure 25a.)

In simpler terms, altitude differences give way to differences in emissivity of the ground formations; this statement is singularly well supported by other aspects. The regions situated 100–140 km to the west are composed of high mountain ridges dominating sunken fault basins, namely Owens Valley, Panamint Valley, Death Valleay, and Amargosa basin: neither tectonic trenches nor mountains are visible, with two exceptions: Grapevine Mountains (37° N, 117° W) and, Panamint Mountain, the southern part which is still snow-capped at this time of year. The latter region, that of Telescope Peak, is not colder, however, than the very broad Darwin Plateau located 60–70 km to the west.

One final example will suffice for this demonstation. In Mexico, the Cerro Pinicate culminates at about 1500 m and is characterized by a radiance temperature of 59 °C (332 K). Five hundred meters lower down, the radiance temperature should be at least 2° higher, but in fact it is 4° lower. The same thermal inversion rule is found

50 km to the north where the Sierra Pinta is hotter than the peripheral regions at a substantially lower altitude. The same applies around Amboy, Little Lake, etc.

It is obvious that the rate of heating starting at sunrise and the importance of emissivity are the essential factors controlling the level reached by radiance temperatures in the middle of the day. During the night, Death Valley is perfectly delimited by the distribution of emitted radiation since, in this case, altimetric and thus temperature differences and the insolation capacity play the essential role, eclipsing that of emissivity. During the day, Death Valley is no longer recognizable, but the Saline Valley is, although the latter tectonic basin is smaller in size and higher up. In this connection, I will recall (Pouquet, 1969, NASA and Rev. Berlin) that the Saline Valley is carpeted with granite debris while Death Valley is covered with sands and salts, with some exceptions which we shall discuss.

(b) *Petrographic influences.* The high 'abnormal' values we found in Mexico, Amboy, Little Lake, etc. are due to volcanic formations. At the extreme north of Death Valley, the surroundings of Ubehebe crater are spattered with volcanic ash, while basalt flows are found between Grapevine Mountains and the Beatty basin. These volcanic formations have high absorption power and their emissivity is shown up by high radiance temperatures.

The various playas in the Mojave Desert contrast with the peripheral relief, which owe their higher temperatures to granitic and paleozoic outcroppings. These playas (dry lakebeds) are frequently moist and sometimes covered with thin films of water, which explains the temperature drops. Here we should note the extent to which pediments and playas are shown up by the radiance temperature distribution.

3.2. Land use

(a) *Irrigated land.* When the volume of irrigation water is sufficient to saturate the soil by midday, the radiance temperatures immediately give very low values; while already-dry or non-irrigated land appears to be very much warmer. We can immediately see, by using the maps in Figures 25a and 25b, how the California Great Valley is divided into two unequal parts, the center and the north (from Tulare 'lake') showing its agricultural richness, and the south, ever in need of precious irrigation water.

The same remarks apply to the entire sunken basin partly occupied by Salton Sea. The carefully-tended crops in the Brawley-El Centro region contrast clearly with the Mexican part of the basin, with the stretches of dunes near Yuma, and the northern part of the trench. The latter region has only a few scattered oases, such as Palm Springs, incapable of offsetting the high radiance temperatures of the barren land.

(b) *Soil erosion.* The reader will observe the long chopped-up strip following the eastern foothills of the Sierra Nevada, the high temperatures of which contrast with the values noted to the east and west. In fact, this discontinuous belt coincides with gold-mines, exploited at the end of the last century: those of the Mother Lode between Mariposa and Nevada City, California. The unruly activities of the prospectors liberated the streamwater, stripped karst topography... leaving the land arid and incapable of supporting crops. Very faithfully, the Nimbus 4 radiometer picked out these famous regions of California, pointing up the excesses that resulted in soil erosion without parallel in the United States.

We also note a number of warm patches inside fairly cold areas touching and inside the coastal chains and the Great Valley, for example 50 or 60 km south of Stockton. These are low hills totally eroded by overly aggressive agricultural methods.

By contrast, the inland delta of the San Joaquin and Sacramento Rivers shows particularly low values as a result of floods controlled by dikes near the California State Capital.

4. Passive Microwaves

Figure 26 presents a map drawn for a 'press conference' at Le Bourget, France (international aeronautical meeting) in June 1973. As far as I know, this is the second map made from passive microwave data; the first is in a French scientific bulletin under the author's signature. These maps are also distinguished from all the others in that they derive from aircraft instrumentation, and because they are entirely drawn 'by hand' without the aid of the computer. To do this, I had to pick up the log listing, point by point. Also, these data are so rich that I had to content myself with a contouring only every 5 K (land) and 10 K (sea).

All of the man-made constructions show up with characteristic low T_{Br}. The San Diego area known as Pacific Beach presents a very interesting transition between the 'tight' concrete downtown area (not on the map) with very low T_{Br}, and the classical suburb with scattered homes lost in the green foliage, much warmer. The contact between two geomorphic units is depicted by the N-S rectilinear pattern, followed also by the freeway between San Diego and Los Angeles (Interstate 5). Note, also, the very cool golf courses (well watered) and warmer peaks without plant cover.

The most interesting facts lie in the Pacific Ocean, where the 'anomalies' are seen with the dotted areas, i.e., areas warmer than they should be. These hot bodies of water are the consequence of the U.S. Navy rejecting excessive quantities of wastes, chemicals, and exhausts. The mud carried down by Californian rivers reaching the sea could explain the phenomenon, but I doubt it, knowing all too well how desperately dry the beds of these 'rivers' are.

This is only a 'try'. Let us hope that, at the time of Spacelab, we will have the opportunity to map, more accurately (with a computer) the multiple results which will be obtained from the satellite.

5. Conclusion

We have seen the vast potential of remote sensing using reflected and emitted infrared; we have merely glimpsed the wider and most promising possibilities of the hyper frequencies. Also, some discrepancies showed up when comparing results

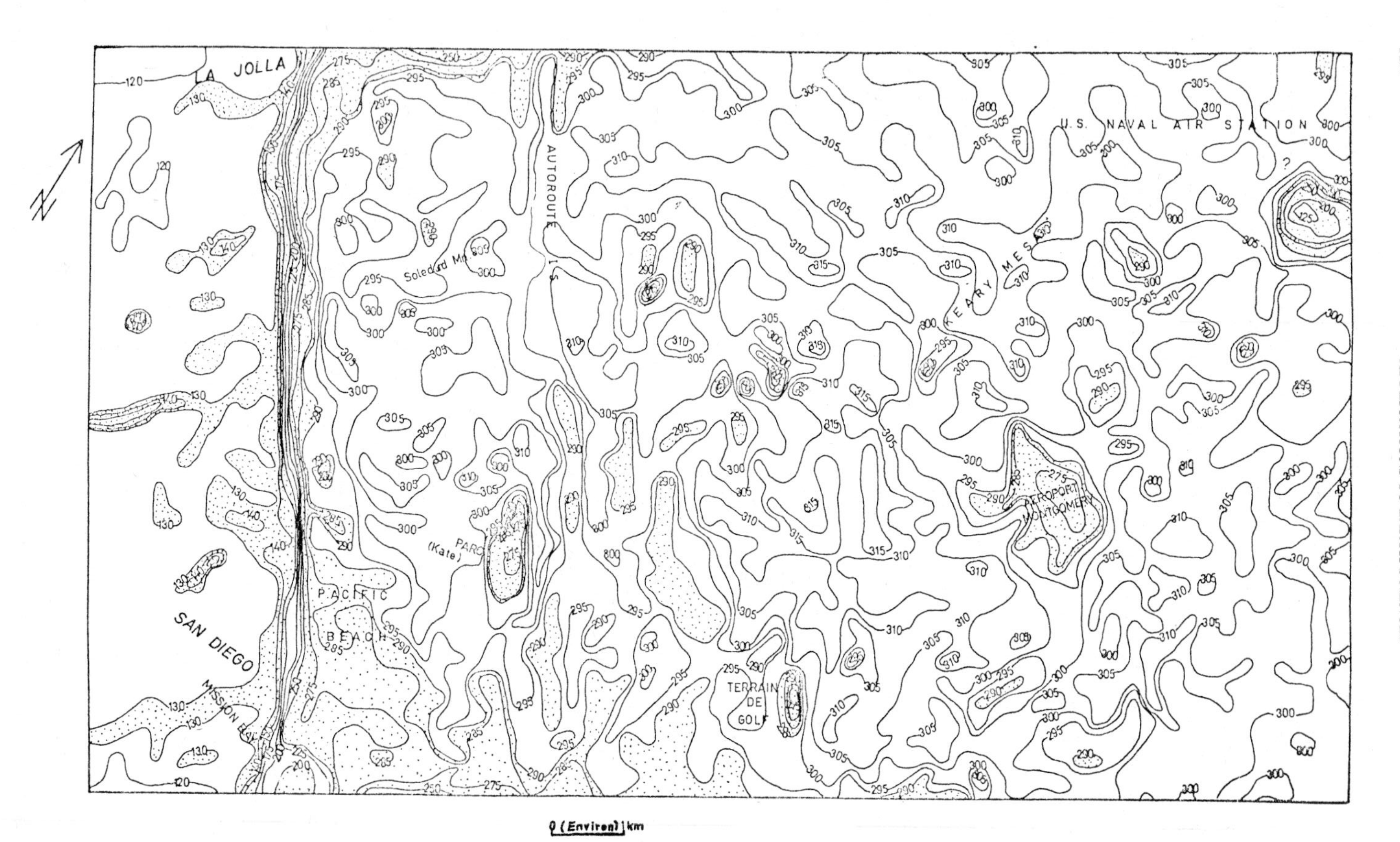

Fig. 26. San Diego area (California) seen through a passive microwave radiometer. – Flight Convair 990, 10 May 1967 between 19:20 hrs (E) and 19:22 hrs (W), GMT. Linear ground resolution: from 125 m (W) to 225 m (E). Passive hyper frequencies 19.35 GHz. Dotted area: thermal anomalies.

obtained in different spectral regions, either at night or during the day. For instance, former river courses do not show up with daytime thermal infrared while they are striking during the night. Very often, too, conclusions reached after scrutinizing such satellite documents were followed by "now we have to check in the field with the classical tools". In fact, just as remote detection is none other than an extra tool available to the earth sciences, so these modern techniques must be operated cooperatively and just not concurrently. These are: (1) systematic examinations of several maps from the same spectral band, (2) permanent comparisons between nocturnal infrared and diurnal infrared (emitted and reflected), and (3) – where possible – massive usage of passive microwaves. These will show up the essential earth science problems.

Having reached this point, only field studies using traditional tools can lead to rational explanations and positive results.

CHAPTER IV

NON-GEOPEDOLOGICAL EXAMPLES

(Volcanology, Glaciology, Oceanography, Meteorology)

As soon as earth sciences in the restricted sense of the term are out of the picture, things become much simpler. For the most part the glaciologist, oceanographer, and meteorologist have little use for the 1:1000000 scale: they are quite satisfied with the 1:5000000 or 1:10000000 scales. On these scales, spurious noise fades into insignificance and satellite wobble has no effect on mapping.

Volcanology, although its rightful place is in the geopedological section, must be included in this chapter since nobody has employed digitized mapping techniques in this subject.

1. Volcanology

Satellites, by their continuous coverage of the globe, provide an unmatched source for detection of continuing volcanic activity or recognition of unmapped regions once affected by such activity. We gave two examples of the latter in Part 3, Chapter I; a further example will suffice to support this statement.

On 19 August, 1966, an eruption took place along an elongated fault in the island of Surtsey in the Vestmannaeyjar Archipelago, south of Iceland. It is estimated that an average volume, 3 to 3.5 m^3, per day of material was expelled between 19 August and 31 December, 1966. This material was an alkaline complex based on olivine basalts, hitting the air at a temperature of about 1130 °C. On 20 August, the lava stream reached the shore.

The eruptions of Kilauea (Hawaii), Etna (Sicily), and Surtsey had been recorded by Nimbus 1 in 1964. In 1966, Nimbus 2 'saw' the eruption, which was being surveyed by aircraft carrying infrared radiometers. Nimbus 2 passed vertically above the region on the second day of the eruption (orbit 1228). On the photofacsimile, the Surtsey 'anomaly' showed up as a single very dark speck being a single scanned sample, namely an area 8 km square. The same observations were made on 22 August (orbit 1315), 8 September (orbit 1541), 16 September (orbit 1648), 20 September (orbit 1701), 21 September (orbit 1774), and 3 October (orbit 1874) (Williams and Friedman, 1970). Unfortunately, we are still waiting for an analysis of the digitized maps.

2. Glaciology

My colleague at Allied Research Associates, John Sissila, demonstrated the ease with which iceberg patterns could be followed using satellites. He mapped the movements of a small ice island about 37 km long using Nimbus 2 television photographs.

Unexpected consequences were found, for instance, the fact that "...the iceberg movements are somewhat abnormal as compared to known currents in the region." The speed of movement was found to average 650 m per hour, with a maximum speed of 1700 m and a minimum speed of 180 m (Sissila, 1969).

Although I know of no practical study, I believe that knowledge of the radiance temperatures of ice could aid navigation in the polar regions: the radiance temperature of ice changes according to its age. In addition, emissivity itself is affected (dust on the ice gives darker colors). Thus, careful study of temperature distribution maps of the icecaps could indicate the best routes for icebreakers, i.e. the higher-temperature regions where the ice is older, probably cracked, thus offering less mechanical resistance.

Figure 27 provides an example of cartography of the northeast quadrant of the Antarctic on 21 November, 1969. The original map was drawn to the 1:5000000 scale. Within the icecap we can distinguish the weaker zones by their lower reflectances; one is in the immediate vicinity of Mackenzie Bay, others are between 20° and

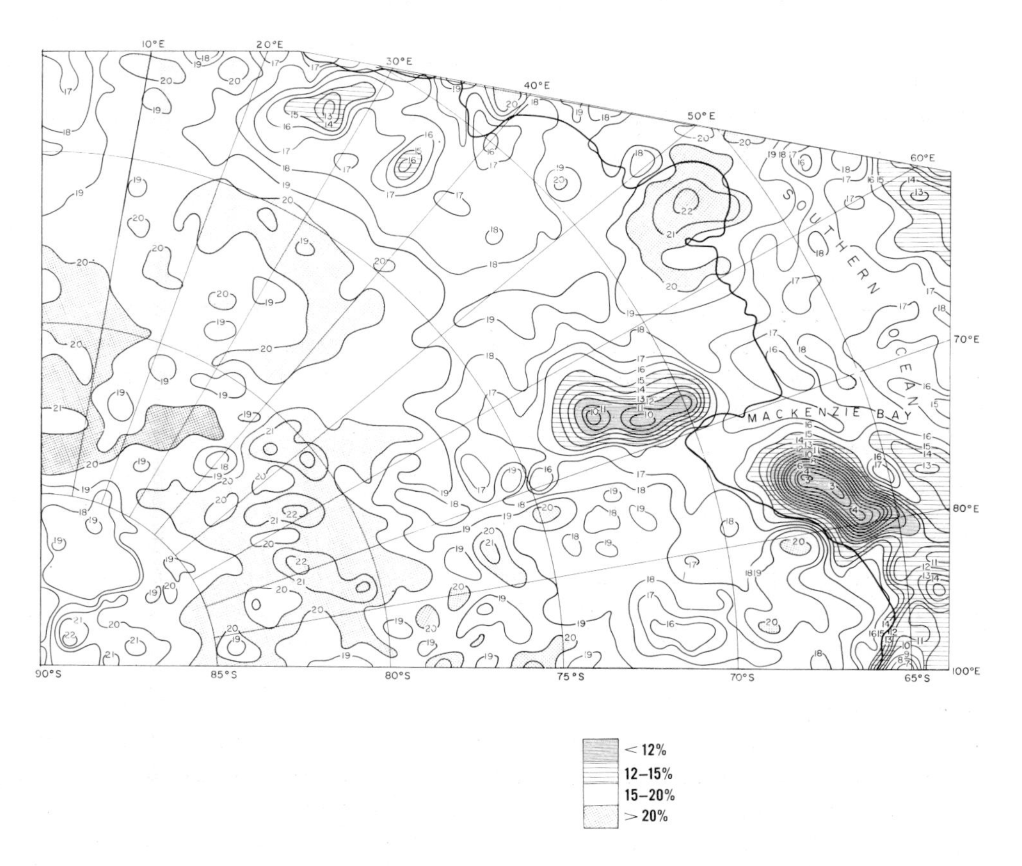

Fig. 27. Infrared albedo, Antarctic region. – Nimbus 3, daytime channel, 0.8–1.3 μm; orbit 2965, 21 November, 1969. Map drawn to the original scale of 1:5000000.

40 °E. Moreover, the 'fresh' ice, or perhaps ice covered with a carpet of recent snow, is shown up by fairly high infrared albedo values, above 20%. Finally, open water is indicated by values below 6–8%, as is the case toward the back of Mackenzie Bay. With the exception of this bay and the shoreline immediately south of it, the icecap extends far beyond the boundaries of the continent.

Since I am neither glaciologist nor oceanographer, I have drawn this map as an example only, leaving correct interpretation to the experts. Ribot (1970) made some studies of the Antarctic using Nimbus 3 data (0.8–1.3 μm): I will consequently refer the reader to his original thesis which I believe to be the first treatment of the subject.

Finally, although I have no example to offer, the invaluable services offered by satellites for continuous surveillance of icefields and glaciers the world over will be readily appreciated. We can see the interest of this knowledge to tourism, hydroelectric facilities, and hydrographic research. In other words, we are in possession of an incomparable instrument for following the *water reserve patterns of our planet*.

3. Oceanography

Many studies have come to light. I will confine myself to three examples; the first has already been partially discussed.

3.1. Water pollution

We have already seen that the presence of chemical pollutants causes a characteristic drop in infrared radiance temperatures, and a steep increase in the microwaves.

Turbidity also alters emissivity (thermal values) and reflectance: both reflectance and radiance temperatures are higher. These principles, as we already know, have helped us to map muddy water patches caused by Nile sediment in the eastern Mediterranean (Pouquet, 1970, Mexico).

It is possible to evaluate the chlorophyll concentration of seawater (Clark *et al.*, 1970). These authors used a spectrometer which, above the Atlantic near George's Bank, showed peaks at 500 mμ and 800 mμ. A cross-check in the sea at the same point showed an excellent relationship between the relative difference in intensity at 500 mμ and 800 mμ and the chlorophyll concentration. These studies were conducted from an aircraft; the only type of satellite capable of such measurements is the ERTS, which carries a spectrometer operating in these spectral bands.

3.2. Ocean currents

Ocean currents are most easily detected from a satellite and literature on the subject is already abundant. Knowledge of these currents is of obvious economic interest and this is one of the reasons why numerous satellite information analyses have been performed on the various Atlantic and Pacific currents. I will cite only the example of the Gulf Stream.

The map in Figure 28 is the first ever to be prepared from satellite data (Warnecke

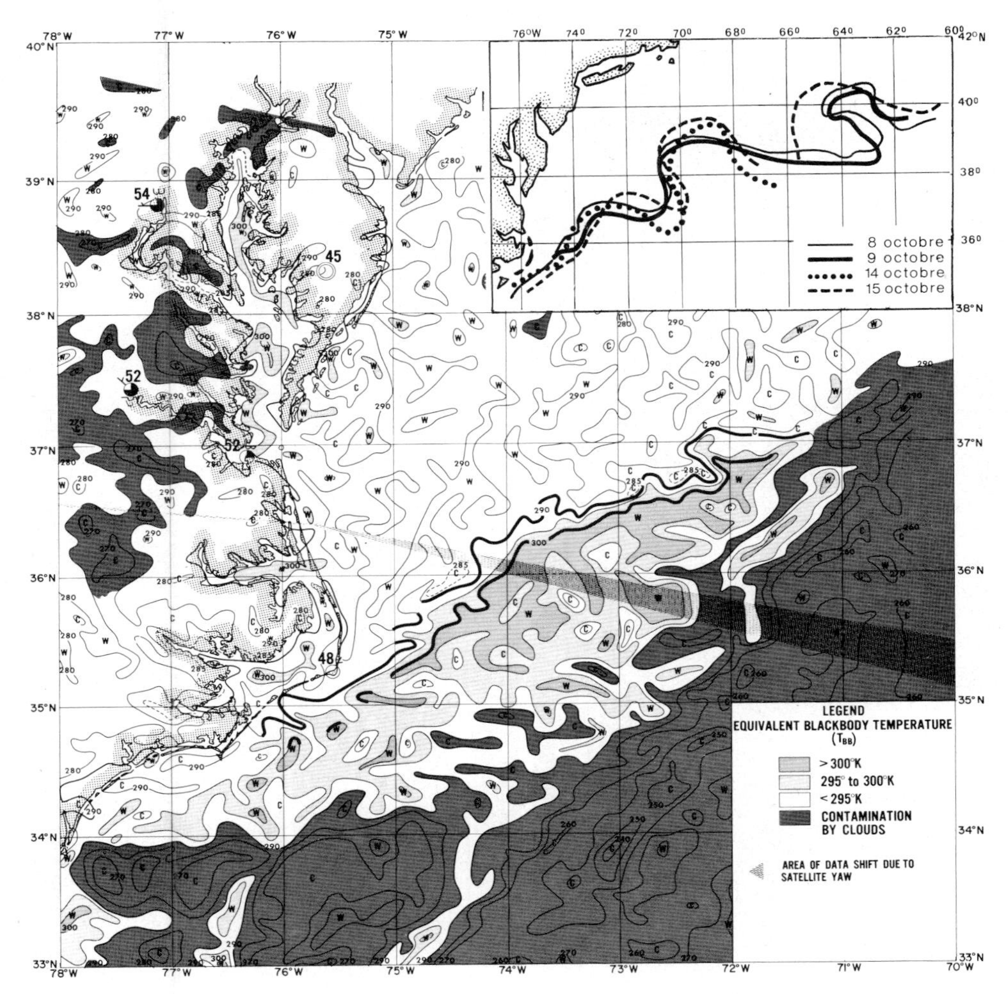

Fig. 28. Northern edge of the Gulf Stream Nimbus 2, nighttime channel, 3.5–4.2 μm, orbits 238–239, 2 June, 1966. *Dark region:* contamination by clouds; *grey region:* transition area, due to satellite movement. Map prepared by G. Warnecke in 1966–1967. *Inset:* Northern edge of the Gulf Stream. Different positions of the current's edge in October 1966 (Greaves *et al.*, 1968).

et al., 1969*). It clearly shows the northern border of the current on 2 June, 1966. The inset shows the successive positions of the same current during October, 1966 (Greaves *et al.*, 1968).

3.3. Movements of Water Masses

Sziekielda (1970) showed how infrared measurements could detect the patterns of large water masses. The example in Figures 29a (28 July, 1966) and 29b (14 September),

* This map appeared in the article cited in the bibliography, but was actually drawn in 1966–1967 by my colleague Gunther Warnecke, Berlin Free University.

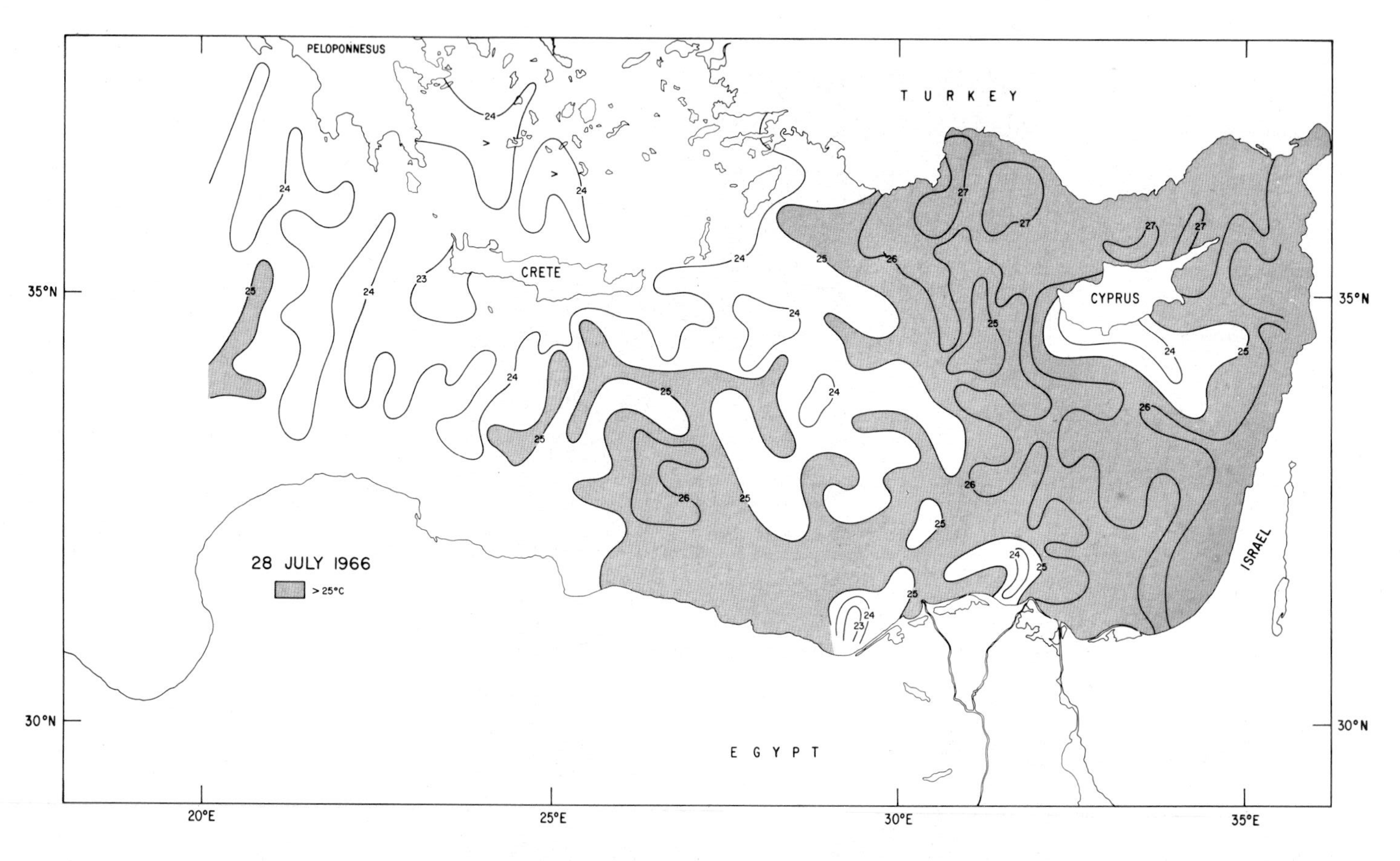

Fig. 29a. Eastern Mediterranean: surface temperature pattern (radiance temperatures) on 28 July 1966. Nimbus 2, nighttime channel, 3.5–4.2 μm. Communicated by K.-H. Sziekielda.

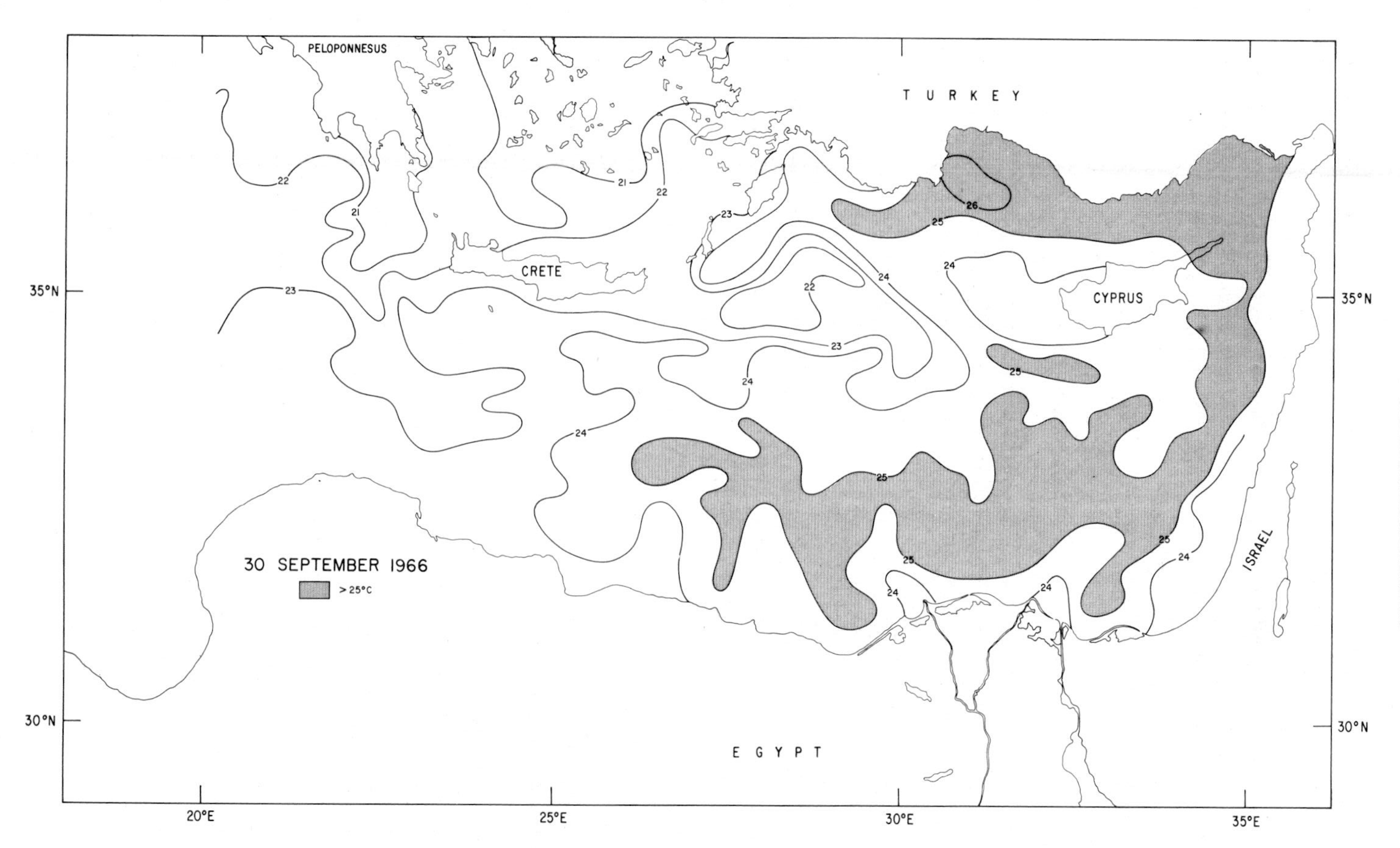

Fig. 29b. Eastern Mediterranean: surface temperature pattern (radiance temperatures) on 30 September 1966. Nimbus 2, nighttime channel, 3.5–4.2 μm. Communicated by K.-H. Sziekielda.

enables us to follow the day-to-day movements of the cold water masses invading the eastern basin of the Mediterranean. In his study of the changing distribution of water masses along the Somali coasts, my colleague showed (with the Nimbus 2 high resolution radiometer) how the southwestern monsoon caused cold water to rise along the shoreline. At the same time, he deciphered the significance of warm and cold water distributions which, incoherent though they may appear, actually evolve similarly to margins of anticyclonic and cyclonic air masses. Only continuous satellite measurements were able to provide these data.

4. Some Meteorological Aspects

These are the most familiar. Most television transmitters, for instance, broadcast a satellite photograph to illustrate the announcer's commentary in the weather forecast. Indeed, research studies are so plentiful that it would be impossible to select a brief bibliography.

The satellites are meteorological, and when they are operational, like the ESSA, they work exclusively on weather-forecasting, using TV and infrared imageries, automatically restored to scale.

Fundamental research is facilitated chiefly by other instruments on board the experimental satellites; when infrared is analyzed, it is in the form of digitized maps, which are not available until about a year later.

4.1. Weather forecasting

Every day, photographic mosaics are made up from all orbits and from television and infrared data, giving a worldwide picture of the distribution of cloud masses, fronts, and atmospheric perturbations. With the ESSA and Nimbus, we have four shots available per 24 h (9:00 and 21:00 for ESSA, 12:00 and 24:00 for Nimbus). Thus, forecasts can be made with an accuracy unknown before the space age.

Another aspect of meteorological observation is shown in photo 2, Plate D, opposite p. 81. Complete global coverage is provided by two stationary satellites, one over the Pacific, the other the Atlantic. Every 30 min, we receive a worldwide picture, considerably facilitating observation of all atmospheric movements. Speeded-up cartoon films have been made of these movements for teaching purposes: for instance, we were able to follow the pattern of the jet stream for the first time on the screen.

Probably one of the most interesting aspects comes from observation of tropical cyclones (hurricanes and typhoons). They are detected at their birth and followed step by step. Speeds, directions, and intensity are immediately evaluated and the coasts likely to be struck are warned in time. Material damage has not been minimized, but I believe the number of victims has been considerably reduced, thanks to the satellites.

4.2. Fundamental research

Photo 2, Plate A, opposite page 59 shows a characteristic aspect of the hurricane

Camille which wrought so much havoc on the American coasts along the Gulf of Mexico in August, 1969. The quantitative study on a digitized map could not be made until a year later: it enabled the experts to expand their knowledge of the mechanisms of such perturbations. Most of the devastating cyclones have been the departure point for similar investigations. In this case, the applications are almost immediate in the sense that, from cyclone to cyclone, they enable better predictions to be made of the structure, speed, and intensity of the phenomena.

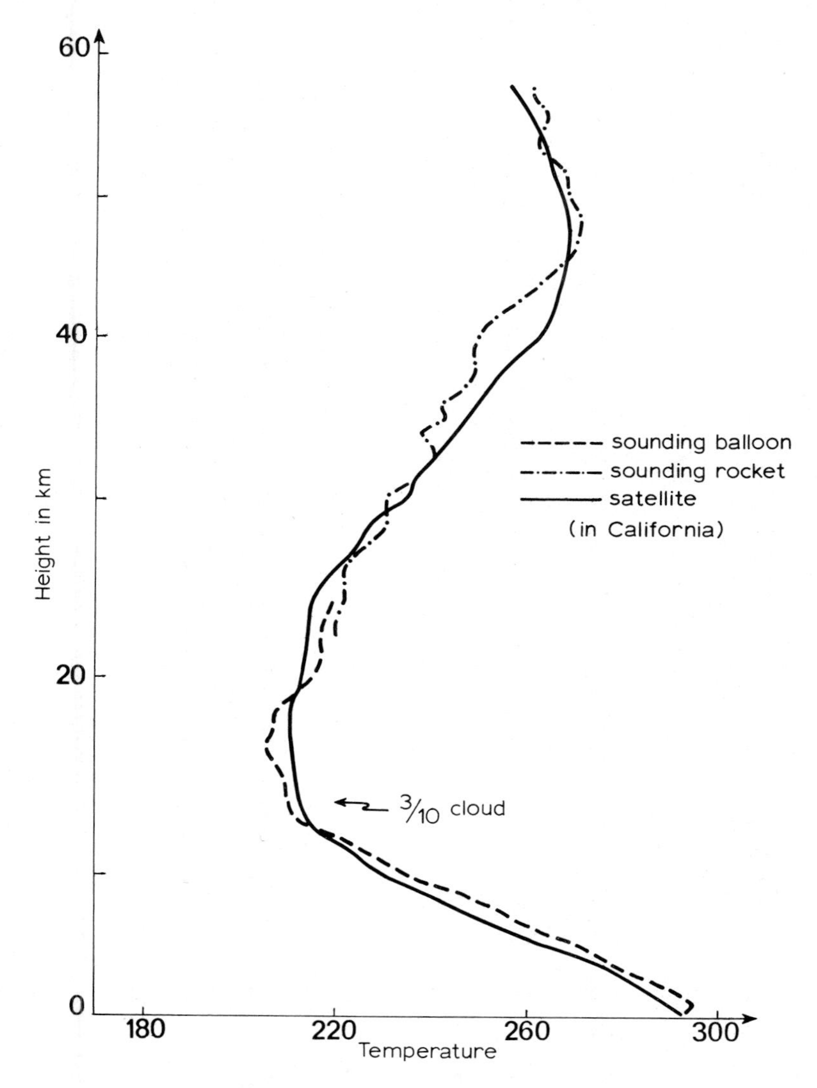

Fig. 30. Temperature profile of the atmosphere and stratosphere derived from Nimbus 4 and radiosonde data on 10 April 1970. Nimbus 4, selective radiometer. Communicated by the 'inventor' of the instrument, 'Jim' Williamson, Oxford University, England.

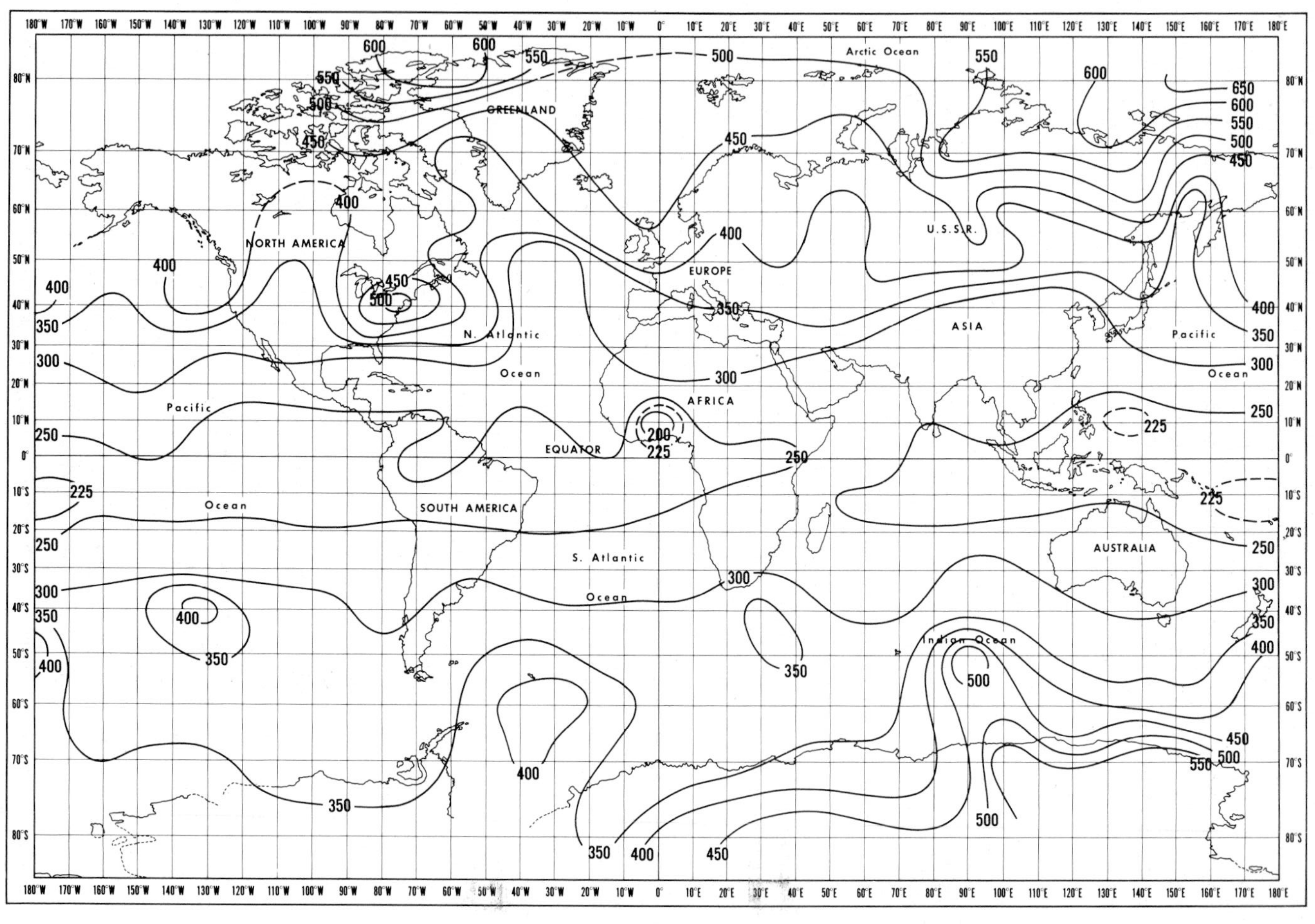

Fig. 31. Ozone distribution, 22 April 1969 (total ozone 10^{-3} cm STP) – From Nimbus 3. Communicated by C. Prashakara, Goddard Space Flight Center.

Two of the participants in my remote detection seminar at San Diego State University (Laporte and McCarter, 1971), using Nimbus 2 cartograms, showed that it was possible not only to determine the cloud type but also to evaluate the activity of various types of weather, intensity of precipitation, wind direction, etc.

Other fields of meteorology, however, have derived the greatest benefit from fundamental research on satellite experiments. The water vapor (6.3 μm), CO_2 (15 μm), and O_3 (UV and 9.6 μm bands) channels have led to progress unsuspected before the satellite era. Photo 1, Plate D, opposite page 81 shows, side by side, the same region seen through (1) the water vapor band, and (2) the atmospheric window band. The relative ease of evaluating the daily fluctuations of atmospheric water content with these digitized maps will be readily understood.

The CO_2 channel has been the source of considerable progress: from this channel we have been able to find out daily, throughout the world, the distributions of and changes in stratospheric temperatures. With Nimbus 4, an additional step has been taken: for the first time we are continuously deriving temperature profiles from ground level to an altitude of 50–60 km. Figure 30 shows the first result obtained two days after the satellite was launched.

Until the satellite age, we had only a rather vague idea of the energy balance of our planet. For example, our knowledge of the albedo (0.2 to 4 μm) was fragmentary and somewhat empirical. Using channel 5 of the medium resolution infrared radiometer (Nimbus, 2, 3, 4, and 5, under another name) we can map this important phenomenon – every day if necessary (although in practice it is done montly).

My colleague C. Prabshakara worked on problems in the ultraviolet and ozone regions. Figure 31 shows one of his maps: we can see the value of this knowledge at a glance.

4.3. Impact on Earth Sciences

A few years ago I presented the results of laboratory experiments suggested by observations in tropical Africa. The most important point was to assign major responsibility to ultraviolet radiation for certain soil genesis phenomena; in particular the massive release of iron and, to a lesser extent, of Ca^{++} and Mg^{++}*. The work of my colleague Prabshakara enabled me to confirm my conclusions in 1970 since, quantitatively and not qualitatively (as before), we can see why these shortwave radiations are affecting intertropical regions. This is only one example among many.

Air and water pollution has become one of the major concerns of our age. I believe that geographers are now in possession of a splendid tool for conducting their investigations. Similarly, climatologists can use satellite records to improve their knowledge, correct numerous errors, and fill in the gaps that existed before the launching of the Nimbus and ESSA satellites. Nimbus 5 and the ERTS enable microclimatology studies and detailed observations to be made, opening the way to all disciplines covered by the term 'geography'.

* J. Pouquet (1965), First results of experimental studies on the release of Fe^{++}, Ca^{++}, and Mg^{++} ions under tropical climatic conditions, *C. R. Som. séances Société Géolog. de Fr.*, 5 April, 1965 session.

PROSPECTS FOR THE FUTURE

We know what progress has been made since the first Sputnik. We are aware of the possibilities offered by modern remote detection techniques. Thus far, we know that these possibilities have been restricted by cartographic problems and the problems of taperecorders on board spacecraft. Last but not least, we can study only a tiny fraction of the planet since infrared is blocked by clouds. It is thus time to take a look at what we might expect from the future.

1. Satellites of the Immediate Future

We can distinguish three stages, the first of which is already underway.

1.1. 'Earth science' series

With Nimbus 5 and ERTS, mapping to scales of 1:50000 and 1:25000 is perfectly possible. Multispectral photography and spectrometry open up new possibilities leading to results that are not only engineering achievements but may well have future applications for the disciplines concerned. In my opinion, development of a microwave radiometer must be considered merely a prelude to a later stage and not an end in itself, since ground resolution is always far too inadequate for earth sciences.

1.2. Series expected in the more distant future

In a more distant future we can hope for technical improvements from the physicists: instruments that penetrate the clouds like microwave but have a resolution like that finally achieved for infrared. We will know we are ahead of the game when the antenna problems have been overcome.

I believe that radar is undesirable onboard a satellite since its viewing angles are inadequate for the ground areas scanned – radar belongs on an aircraft. Moreover, I think there is little use in developing spatial resolutions finer than 0.1 mrad – feasible no doubt, but encroaching on the domain of the airplane which alone is qualified for detailed studies.

Finally, I hope that we will shortly be able to rid ourselves of the burden of onboard recording with magnetic tapes. Taperecorders have all proved disappointing. They introduce new spurious noise, which increases as they deteriorate, and cuts down the useful life of the satellites. A broad international agreement could lead to direct broadcasting of all information gathered by onboard instruments to stations all over the world. Each of these receiving centers would immediately retransmit all data to a central agency.

1.3. Internationalizing operational satellites

The heading uses the term 'operational' as opposed to 'experimental'. Experimental studies must be pursued, as in the past, under the aegis of local institutions such as NASA and ESRO*, to mention only the U.S. and Europe. International collaboration in this field can only continue to benefit fundamental research.

The picture is different when we look at operational satellites, actively *surveying* crops, forests, oceans, rivers, industrial plants, traffic, etc. throughout the world. Everyone is his own master at home, and nobody likes his neighbor peeping over the fence.

The Sputniks are, and must stay, *Soviet*; the Nimbus are, and must stay, *American*; the Fr., A, D, are, and must stay, *French*...; the ERTS are but should *not* be American. In practice, one can hardly view the Earth Resources satellites as experimental even if the scientific payload (deliberately) operates only over a predetermined region. With the ERTS we are entering the operational phase, and no-one can hold down an exclusive on them.

In actual fact, and I speak from a position of knowledge, the Nimbus like the ERTS are the fruit of international scientific collaboration in which the Americans, the British, the French, the Germans, and the Italians, to cite only the principal investigators in alphabetical order, supplied basic scientific knowledge. The Americans contributed to success by a material organization unmatched in Western Europe and by their predilection for risk, which happily offsets the European traditional caution.

To leave information derived from satellites to the discretion of a single power would be a profound mistake. I believe that only an international service, reporting to the United Nations and more precisely to UNESCO, would be qualified to build, launch, and operate the Earth Resources satellites, and alone to manage all information and furnish it, on application, to the countries concerned.

The necessary international collaboration took shape at Brussels, in July 1973. The European nations agreed between themselves on the NASA-ESRO post Apollo program.

Actually, a 'sequence' of compromises helped toward the final decision, compromises taking the form of "if you give me some salt, I'll give you some pepper". France agreed to pay her share for Spacelab on the condition that Great Britain and Germany participate in the French project of L3S (booster). On the other hand Germany assisted with the L3S booster in return for the French help for Spacelab. Great Britain was mainly concerned with her Marots satellite project for maritime navigation, France and Germany sharing a part for this British satellite. Great Britain agreed to participate in Spacelab and L3S.... Finally, the other European nations, Italy, The Netherlands, Belgium, Scandinavia etc. entered the 'game'.

The door is open and the future looks bright....

* ESRO changed its name in the summer of 1974, in ESA, European Space Agency.

2. The True Future of Remote Detection

Here I will be brief, as all of these ideas have been discussed before, particularly in the conclusion to Part 2.

If you want to drive a nail into a plank or pierce a sheet of paper with a thumbtack, you *could* use a sledge-hammer, but it is more sensible and economical to leave to the sledge-hammer what the humble tool cannot do, and vice versa.

This is exactly the satellite-airplane problem. I have said it before and I will say it again: ground resolution of less than 100 m is not the province of the satellite. It costs too much. The airplane will do the work more cheaply and much better. The real future of remote detection, excepting the improvements mentioned above, will be a partnership between the satellite and the airplane, each one staying in its respective domain. What is more, the coming generation – the students – must be prepared for this real future.

3. University Aspects

The future is conditioned by the University. As long as the university confines itself to relishing the fruits of the past, satellites and aircraft equipped with sophisticated instrumentation will be nothing but costly playthings, prestigious – but useless.

Not long ago, a pessimistic book was published by Henri Wadier*. I fear that this author is the type of person who, looking at a still half-full bottle, considers it sadly to be already half-empty. The modern university, and this applies to every country in the world, is undergoing readjustment since the gap between technical progress and our faculties of assimilation has increased out of all proportion. It would be astounding if we could not, for the first time in man's history, learn to readjust.

I have every confidence in the future. Our universities still possess most of their scientific value and prestige. I am convinced that the coming generations, the powerful users of space techniques, will be decently trained. These prestigious toys, the airplane and the satellite, will become the instruments of choice in combination with our traditional working tools.

* H. Wadier, *La réforme de l'enseignement n'aura pas lieu* [The Reform of Education will not take place], Laffont, Paris, 1970.

APPENDIXES

APPENDIX A

THE GREAT FIRSTS OF THE SPACE AGE

(USA = U.S. Army; USN = U.S. Navy; USAF = U.S. Air Force)

Launch date	Country or agency	Name of Satellite	Chief events
4 Oct. 1957	U.S.S.R.	Sputnik 1	First artificial satellite.
31 Jan. 1958	U.S.A.	Explorer 1	First American satellite.
		Tiros 1	Discovery of Van Allen belt.
12 Sept. 1959	U.S.S.R.	Luna 2	First touchdown on Moon.
1 Apr. 1960	NASA		First photograph of the Earth.
12 Aug. 1960	NASA	Echo 1	First relay satellite (voice, TV pictures).
19 Aug. 1960	U.S.S.R.	Sputnik 5	Dogs placed in orbit, recovered on 18th orbit (Belka and Strelka).
4 Oct. 1960	U.S.A.	Courier 1B	First active relay satellite (telecommunications).
12 Apr. 1961	U.S.S.R.	Vostok 1	First manned satellite (Yuri Gagarin, one orbit).
29 Nov. 1961	U.S.A.	Mercury-Atlas 5	Chimpanzee Enos recovered after two orbits.
20 Feb. 1962	NASA	Mercury-Atlas 6	First American astronaut (John Glenn, 3 orbits).
26 Apr. 1962	U.S.A.	Ariel 1	First U.S.-Britain cooperation satellite.
28 Sept. 1962	Canada	Alouette 1	First Canadian satellite. Ionosphere studies.
28 Aug. 1964	NASA	Nimbus 1	Television, HRIR.
12 Oct. 1964	U.S.S.R.	Voskhod	Three astronauts. 16 orbits.
17 Feb. 1965	NASA	Ranger 8	First close-up photographs of the Moon.
9 March 1965	USN USA USAF	******	8 different satellites launched by the same rocket, Thor-Agena D.
18 March 1965	U.S.S.R.	Voskhod 2	First space walk (A. Leonov).
23 March 1965	NASA	Gemini 3	First attempt to maneuver satellite in space (V. Grisson and J. Young, three orbits).
26 Nov. 1965	France	A.1	First French satellite. Transmits for two days.
3 Feb. 1966	ESSA	ESSA 1	First operational meteorological satellite.
15 May 1966	NASA	Nimbus 2	TV, HRIR, MRIR.
12 Sept. 1966	NASA	Gemini 11	First space docking.
6 Dec. 1966	NASA	ATS 1	Geosynchronous satellite.
15 Feb. 1967	France	D 1D	Use of laser; Doppler effect.
17 Apr. 1967	NASA	Surveyor 3	Soft landing on Moon. Photographs. Soils.
23 Apr. 1967	U.S.S.R.	Soyuz 1	First space victim: V. Komarov whose parachute did not open properly.
5 May 1967	G.B.	Ariel 3	First wholly British satellite.
18 Oct. 1967	NASA	OSO 4	First pictures of Sun in far ultraviolet.
27 Oct. 1967	U.S.S.R.	Kosmos 186	First automatic docking (with Kosmos 188).
7 Nov. 1967	NASA	Surveyor 6	Moon landing. First rocket launched from Moon.
15 Sept. 1967	U.S.S.R.	Zond 5	First lunar orbit. Zond 5 recovered from Indian Ocean.

Appendix A (Continued)

Launch date	Country or agency	Name of satellite	Chief events
11 Oct. 1968	NASA	Apollo 7	Three astronauts made the first rehearsal prior to landing on the Moon.
14 Jan. 1969	U.S.S.R.	Soyuz 4 and 5	First rendezvous of two manned satellites. First first-aid test in space.
3 March 1969	NASA	Apollo 9	First maneuver with lunar vehicle.
14 Apr. 1969	NASA	Nimbus 3	TV, HRIR, MRIR, UV, reflected IR.
18 May 1969	NASA	Apollo 10	Lunar orbits. First color TV pictures.
16 July 1969	NASA	Apollo 11	Moon landing July 20 (Armstrong and Aldrin).
11 Oct. 1969 12 Oct. 1969 13 Oct. 1969	U.S.S.R.	Soyuz 6, 7, and 8	First launching of 3 manned satellites. Vacuum welding experiments. Automatic navigation experiments. 7 astronauts.
8 Nov. 1969	Germany	Azur 1	First German satellite.
8 Apr. 1970	NASA	Nimbus 4	High resolution IR radiometer with 11 μ window.
10 March 1970	Germany	Dial	Launched by French rocket Diamant B.
12 Dec. 1970	France	Peole	Geodesic experiments. Synchronous experiments with meteorological balloons.
15 Apr. 1971	France	D 2A	Hydrogen study in geocorona of Sun, interplanetary space.
16 Aug. 1971	France	Eole	Meteorology (fundamental research) in liaison with balloons reaching about 12 km altitude in the southern hemisphere.
28 Oct. 1972	Great Britain	Prospero	First scientific satellite launched by Great Britain's own booster.
31 Jan. 1972	ESRO	Heos 2	International scientific satellite.
14 Febr. 1972	U.S.S.R.	Luna 20	Lands on the Moon 21 Feb. Returns to Earth 25 Feb.
4 Apr. 1972	France	SRET 1	Launched piggyback with the U.S.S.R. communication satellite Molniya 1.
16 Apr. 1972	U.S.A.	Apollo 16	Fifth successful landing on the Moon 20 Apr.
23 July 1972	U.S.A.	ERTS 1	First satellite devoted to Earth resources.
Dec. 1972	U.S.A.	Nimbus 5	First time a passive microwave operates aboard a satellite. First time a very high resolution thermal infrared operates aboard a satellite.
June–Nov. 1973	U.S.A.	Skylab 1, 2 and 3	First orbital laboratory. First 'big repairs' performed in space. First animal born in Space (a fish). First time a spider spun a web in space.

The California periodical TRW Space Log regularly publishes the latest satellite launchings, an exhaustive list that includes failures. Refer to 'Winter 19..–19..' numbers.

APPENDIX B

WHERE TO OBTAIN SATELLITE DATA

1. REQUEST FOR PHOTOGRAPHS (positives and negatives)

(a) NASA Goddard Space Flight Center, Data Utilization Center, and/or National Space Data Center, Greenbelt, Maryland 20771.
HRIR (THIR for Nimbus 4 and 5). For Nimbus 3 through 5, specify Night or Day.
MRIR: specify channel (1, 2, etc.).

(b) National Environmental Satellite Center (ESSA), Suitland, Maryland, satellites ESSA 3, 5, 7, 9 (television).

(c) National Weather Records Center, ESSA, Asheville, N. Carolina. ATS 1 and 3, television plates of Nimbus, TIROS, and ESSA I.

2. MAGNETIC TAPES (d = digitized, r = raw)

NASA Goddard Space Flight Center, National Space Data Center, Greenbelt, Maryland.
Television from Nimbus 3 (r), HRIR (d), MRIR (d), Nimbus 1, 2, 3, 4 and 5 (THIR), IRIS, SIRS, MUSE from Nimbus 3 (d); MRIR from TIROS 2 to 7 (d).

3. COMPUTER-PRODUCED DIGITIZED MAPS

Same address as above. HRIR (THIR for Nimbus 4 and 5, MRIR for Nimbus and MRIR for TIROS 2, 3, 4, and 7 (See Appendix D).

4. SPECIAL RECORDS, BY REQUEST
(a) National Weather Records Center (ESSA), Asheville, N. Carolina: microfilms of cloud analyses (TIROS, ESSA).

(b) NASA, Goddard Space Flight Center, National Space Data Center, Greenbelt, Maryland. Statistical lists of digitized data on HRIR and MRIR, Nimbus. By exception, microdensitometry.

5. PHOTOGRAPHS TAKEN BY GEMINI AND APOLLO ASTRONAUTS
(a) NASA Manned Spacecraft Center (MSC), Earth Resources Program, Houston, Texas.

(b) Technology Application Center, University of New Mexico, Albuquerque, New Mexico.

REQUESTS FOR PHOTOGRAPHS (FACSIMILE, TELEVISION)

Using catalogs, specify name and number of satellite, orbit number, and day. It is not necessary to state longitude, but give the latitude to the nearest 10° (ending numbers with 0 (10°, 20°, 40°, etc.) unless the entire orbit is required (in this case you will receive contact prints). For meteorology the above information will suffice; if interested in land and sea, add to application 'dodged for terrestrial features.'

Give all necessary information for format desired: *contact* (70 mm films); specify positive, negative; on transparent film, on paper. For *enlargements*, keep to 8″ × 10″ format. Avoid asking for larger sizes.

I strongly advise asking only for *negatives on 70 mm film* since these are easier to process, in the local photography lab if necessary.

Photographs applied for from ESSA come from 35 mm negatives ('24 × 36 mm'). In this case, it is preferable to ask for a negative copy in the same format and make transparencies or enlargements in your own laboratory.

APPENDIX D

APPLICATION FOR DIGITIZED MAP

Request: Computer produced grid print map (digitized data).

(A) From: ______________________ (B) Organization ____________ *Mailing address* ______________________ (C) *Purpose* ______________________

______________________ (D) *Date* ______________________

(E) *Satellite* ______________________ (F) *Channel* ____________ (G) *Day or night* __________

(H) Projection: ______________________ (I) *Scale:* 1: ______________________

(J) *Contouring:* every __________ K __________%

Begins at __________ K.

If infrared albedo (reflectance), add:

Multiplier: 1.00000

Base: 10. *Contouring* – (every __________%)

Maximum nadir angle __________°. If possible, *filtered data.*

D/O	*R/O*	*Date*	*Calendar day*	*Time*		*Latitude*		*Right West Longitude*	*Deg. of long. per mesh inter.*	*Nr of mesh per degree of long.*	*Total number of mesh*
				Begins	*Ends*	*Hi.*	*Low.*				
1	2	3	4	5	6	7		8	9	10	11

(Do not write the letters (A), (B), nor figures 1, 2, 3)

Explanations

(A): Professor, Dr, Research Dir., etc. and name.

(B): University, Department, Institute, Research Center, etc. and mailing addresses.

(C): (Reasons for application.) The following formula is usually the most appropriate with specific explanations: 'Research performed by Faculty members and advanced students under the leadership of professor...'

(D): Date of application (day, month, year).

(E): Name of satellite and number (print clearly, in boldface type).

(F): Detector used, HRIR, MRIR, THIR, etc. For THIR, specify channel (11.5 or 6.7 μ).

(G): Specify whether day or night information is required, especially for Nimbus 3 HRIR and Nimbus 4 and 5 THIR, SCMR. MSS channel for ERTS (only channel 4, 5, 6, 7).

(H): Map projection: Mercator only or, for polar regions, polar stereographic projection.

(I): Scale. Largest is 1:1 000 000 (Mercator projection) or 1:5 000 000 (polar stereographic projection). See footnote, p.146).

(J): Starting point for equal value curves (heat, reflectance). For radiance temperature maps, it is better to keep to intervals of 2 K. For meteorological studies ask for 5°, which is quite sufficient.

For nighttime, do not start before 270 K, 280 K being a good beginning for day values. These beginnings depend on the season and region involved.

For infrared albedo (reflectance) it is wise not to write merely 'every 1%'. In this case, indicate Base 10 (punched on cards) and contouring ______. For 1% write −10; for 20% write −20, for 0.5% write −05, etc.

Nadir angle, i.e. angle formed by the vertical and the boundaries of the region in question. Never go beyond 50°, the extreme limit; usually keep to 40° or less.

Table Columns

1. *D/O*: Data Orbit number, which always appears on the catalogs.

2. *R/O*: Readout Orbit, in which the satellite was interrogated. In general, but not always, this information appear in the catalogs. If this number is unknown leave a blank as this datum, together with 3. below, is not used to punch the cards but only to more easily locate the magnetic tape needed.

3. *Date of orbit*, e.g. 23 June ... (day, month, year).

4. *Calendar day, very important*: the number between 1 and 365 (or 366). For example, 27 February = 58; 13 July = 194 (or 195), etc.

5 and 6. Time satellite pass begins and ends. Show GMT hour and minute at which satellite entered and left the region. (See the paragraph *Determination of Pass Time* below).

7. Indicate highest latitude (Hi) and lowest latitude (Low), using the + sign for north latitudes and the − sign for south latitudes. (Do not write N or S.)

8. Indicate longitude of the *right boundary* of the region, i.e. the east longitude, *exclusively in west values*. If this longitude is 28° E, write 332, without W (or E). If 28 (with or without E) is written in error, the map will show only blanks.

9. This is the number of degrees longitude between two crosses, and gives the scale. The paper is standardized and the distance between two crosses (one mesh) is always half an inch whatever the scale. On the scale of 1:1 000 000, one mesh covers 0.125° longitude. On the 1:5 000 000 scale, one mesh covers 0.625° longitude, etc.

In this column, write 0.125 for 1:1 000 000, 0.250 for the 1:2 000 000, etc. (In practice one omits the 0, writing merely .125).

Distances in Centimeters between Successive Degrees of Latitude

For a Mercator projection on the 1:1000000 scale, on the paper used for IBM printout

Latitude	Distance	Latitude	Distance	Latitude	Distance	Latitude	Distance
0°		*20°*		*40°*		*60°*	
——	10.13 cm	——	10.82 cm	——	13.35 cm	——	20.68 cm
1°		21°		41°		61°	
——	10.14	——	10.89	——	13.55	——	21.34
2°		22°		42°		62°	
——	10.15	——	10.97	——	13.78	——	22.05
3°		23°		43°		63°	
——	10.16	——	11.06	——	14.02	——	22.81
4°		24°		44°		64°	
——	10.17	——	11.15	——	14.26	——	23.64
5°		25°		45°		65°	
——	10.18	——	11.23	——	14.50	——	24.55
6°		26°		46°		66°	
——	10.19	——	11.34	——	14.77	——	25.55
7°		27°		47°		67°	
——	10.21	——	11.44	——	15.04	——	26.64
8°		28°		48°		68°	
——	10.23	——	11.54	——	15.34	——	27.82
9°		29°		49°		69°	
——	10.26	——	11.65	——	15.64	——	29.11
10°		*30°*		*50°*		*70°*	
——	10.30	——	11.78	——	15.96	——	30.54
11°		31°		51°		71°	
——	10.33	——	11.91	——	16.30	——	32.14
12°		32°		52°		72°	
——	10.37	——	12.03	——	16.67	——	33.92
13°		33°		53°		73°	
——	10.41	——	12.17	——	17.06	——	35.94
14°		34°		54°		74°	
——	10.46	——	12.31	——	17.48	——	38.22
15°		35°		55°		75°	
——	10.51	——	12.48	——	17.92	——	40.81
16°		36°		56°		76°	
——	10.56	——	12.64	——	18.40	——	43.79
17°		37°		57°		77°	
——	10.63	——	12.80	——	18.92	——	47.21
18°		38°		58°		78°	
——	10.69	——	12.98	——	19.48	——	51.33
19°		39°		59°		79°	
——	10.75	——	13.16	——	20.08	——	56.21
20°		*40°*		*60°*		*80°*	

10. Number of meshes per 1° longitude. This confirms the scale. There are 8 meshes for the 1:1000000 scale; 4 for the 1:2000000; 16 for 10° on the scale of 1:5000000, etc.

11. *Total number of meshes*: indicate longitudinal extension. It is futile (and ill-advised) to write the longitude on the left of the map, which could cause the operators to make errors. On the 1:1000000 scale, if the region studied extends over 12° longitude, write 96 in this column (8 meshes per degree).

NEVER WRITE 'between *X* degrees and *Y* degrees E (or W) which could cause numerous errors in card punching (there is not enough personnel to make the necessary corrections)*.

DETERMINING PASS TIMES

Refer to the catalogs giving the pass time over the equator. Read descending node for night information (N) and ascending node for the day (D) except for ERTS (exactly the reverse). Remember that in the day (D) the satellite is northbound, so read in ascending node, and vice versa for the night. Use the time scale that comes with the Nimbus 3 catalogs. For Nimbus 4 and 5, the time scale is shown on each side the daily compilations. For caution's sake it is preferable to start a minute earlier and end a minute later. Count roughly 0.3 min of an hour for 1 degree of latitude.

* For ERTS, channel 7 (i.e. channel 4 of MSS) correct as follows: scale 1:25000; 6 meshes per minute of longitude; 10 s of longitude per mesh. For channel 8 (MSS channel 5) write, in succession, 1:50000; 3 and 20.

For Nimbus 5 and F, change as follows: SCMR; 1:100000; 3 meshes for 2 min of longitude; 40 s of longitude per mesh.

Remember that the scales are approximate.

APPENDIX E

NASA QUESTIONNAIRE ON UTILIZATION OF ERTS (EARTH RESOURCES SATELLITES) AND SKYLAB INFORMATION

After a number of preparatory meetings in 1969 and 1970, NASA published on 2 February, 1971 a document important to workers in all disciplines. The summary page at the end of this section should appear at the top of all applications for participation: p. 6 of "Preparation of proposal for investigation using data from *Earth Resources Technology Satellite (ERTS 1)* and *Skylab (EREP)*."

The questionnaire below is not for evaluating the merits of proposals. In fact, the research program is to be widely developed, explained, digitized, and localized according to this (very simplified) organic plan:

I. SCIENTIFIC PERSONNEL. Names, titles, qualifications.

II. SCIENTIFIC AND TECHNICAL DESCRIPTION OF PROJECT.

(a) *Previous work* (in the same field). Publications, where? when?...

(b) *Objectives*, Examples: checking scientific or technical hypotheses; applications....

(c) *Method and Means of Approach*. Exhaustive description of scientific and technical methods to be used (photogrammetry, densitometry, holography, multispectral projections (see Part 1, Chapter II and Figure 4), programming, etc.

(d) *Results Expected* (obviously the shortest section).

(e) *Projects of principal investigator that involve information from NASA*.

(1) General introduction.

(2) Utilization of information. How? Discussion on instruments to be used.

(3) Photography (techniques, parameters).

(4) Results expected ('enhanced' photographs, charts, diagrams).

(f) *Information needs*: types of ERTS and Skylab records. Needs for aerial flyovers, field experiments, etc. This part must be written in accordance with Appendices B, C, and D and, for Skylab, in accordance with the *MSC EREP User's Handbook* published in February, 1971.

III. MATERIAL AND FINANCIAL NEEDS: offices, laboratories, scientific equipment (description, prices, suppliers, etc.)

Technical Personnel.

Resources (government or other origin). Aerial flyovers, etc.

Expenses must be assessed as accurately as possible under each heading.

IV. CONTRACTUAL NEEDS, exclusively according to Appendix E.

V. HOW TO ASK FOR FINANCIAL AID, see Appendix F.

These appendices, much too detailed for gross oversimplification here, will be found in *Preparation of Proposals* cited above.

Special attention (not necessarily in this order) will be paid to multidiscipline projects concerning the disciplines enumerated in the Summary Proposal Form. For example, see below the more detailed list for Geography/Demography/Cartography and Geology (Appendix A of document published 2 February, 1971).

GEOGRAPHY/DEMOGRAPHY/CARTOGRAPHY. Classification of used and developing lands. City planning. Rural development. Population densities, locations, patterns. Archaeology, Inventory of natural disasters. Inland traffic systems. Orthographic cartography. Orthophotographic cartography. Thematic cartography. Climatological cartography.

GEOLOGY. Minerals exploration. Oil exploration. Environmental hazards. Volcanos. Landslides. Analysis of land formations (geomorphic units; lithological units; structural units).

Similar details are supplied for each of the broad disciplines (agriculture, forestry studies, hydrology, oceanography, meteorology, environment quality/ecology, development of interpretation techniques, instrument technology). I will repeat here the last sentence of Appendix A: *"The principal investigator must identify his research projects based on the disciplines in this list."*

Here are recalled the principal parameters for 'ERTS 1' and ERTS B.

Orbiting altitude: 911.83 km. Inclination to equatorial plane: 99.088°. Period: 103′16″. Excentricity: 0. Equator pass time (local solar time): 9:30 and 21:30. Globe coverage: in 18 days (251 orbits). W–E breadth of land strips covered effectively in a single orbit: 180 km.

INSTRUMENTATION: Television (RBV): camera 1, 480–575 mμ (green band); camera 2, 580–680 mμ (red); camera 3, 690–830 mμ (near infrared). (Known as channels 1, 2 and 3). Unfortunately, the RBV system is out of order.

Multispectral spectrometer (MSS): channel 1, 0.5–0.6 μm (green); channel 2, 0.6–0.7 μm (red); channel 3, 0.7–0.8 μm (very near infrared); channel 4, 0.8–1.1 μm (near infrared); channel 5 (ERTS B only), 10.4–12.6 μm (atmospheric window, thermal infrared). (All channel known as channels 4 through 8.)

Records are supplied to users mainly as 24 × 24 cm photographs, image area is 18.5 × 18.5 cm covering slightly over 1°30′ latitude and longitude, at a scale of 1:1 000 000. They come mainly in 'bulk' (without marginal indications) or more rarely, 'precision image annotation' (with coordinates and main parameters). These proofs are sent on application either as negatives, paper positives, or transparency positives,

Summary Proposal Form

PRINCIPAL INVESTIGATOR (PI) ______________________

ORGANIZATION/ADDRESS ______________________

PHONE NUMBER ______________________

INVESTIGATION DISCIPLINE(S) (LIST ALL IN ORDER OF IMPORTANCE)

e.g., AGRICULTURE, FORESTRY, GEOGRAPHY / DEMOGRAPHY / CARTOGRAPHY, GEOLOGY, METEOROLOGY, HYDROLOGY, OCEANOGRAPHY, ENVIRONMENTAL QUALITY / ECOLOGY, INTERPRETATION TECHNIQUES DEVELOPMENT, SENSOR TECHNOLOGY).

____
____
____
____
____

PURPOSE OF INVESTIGATION (BRIEF DESCRIPTION) ______________________

____

GEOGRAPHICAL LOCATION ______________________

SPACECRAFT (*ERTS* / *SKYLAB* / BOTH) ______________________

FUNDS REQUIRED (YES/NO) ______________________

NASA FUNDS REQUIRED $ ______________________

OTHER FUNDS REQUIRED $ ______________ SOURCE(S) ______________

AIRCRAFT DATA FLIGHTS REQUIRED (YES/NO) ______________________

SOURCE(S) (NASA, DOD, CONTRACTOR) ______________________

GROUND TRUTH REQUIRED (YES/NO) ______________________

TEST SITE LOCATION ______________________

NEW INSTRUMENTATION REQUIRED ______________________

NEW INSTRUMENTATION TO BE PROVIDED BY ______________________

IS AUTOMATIC DATA PROCESSING EQUIPMENT (ADPE) REQUIRED (YES/NO) __________

____

OTHER CAPITAL EQUIPMENT REQUIRED (YES/NO) ______________________

DURATION OF INVESTIGATION (MONTHS) ______________________

FORMAT OF DATA ______________________

or in composite colors. It is of course, strongly recommended to ask also for digitized data (densitometry, effective radiance, reflectance, and radiance temperatures for ERTS B channel 5 (NASA, 1971, *Image Formats and Annotation*).

USEFUL ADDRESSES

(1) REQUESTS FOR INFORMATION (documentation, technical notes, etc.). Contact: Doctor Arch B. Park, Office of Space Science and Applications, Earth Observations Program, code SRR, NASA H.Q., Washington, D.C. 20546, and Mr. John D. Koutsandreas at the same address (code SRB).

(2) PROPOSALS FOR MULTIDISCIPLINE RESEARCH based on *ERTS* and/or Skylab (for non-U.S. agencies):

Office of International Affairs, Code I, NASA H.Q., FOB 6, Washington, D.C. 20546.

Only applications transmitted through government space agencies will be taken into consideration.

EMITTED ENERGY AND EMISSIVITY

The purpose of this appendix is to facilitate the task of users of modern remote detection techniques, and especially to help them avoid the errors frequently committed as to the significance of *radiance temperatures*.

Figure 32 illustrates Plank's law. We observe three groups of curves between 200 K (−73 °C) and 440 K (167 °C):

(a) *Long wavelengths*, from 15 μm to 50 μm, characterized by low energy emission (spectral radiance) which increases rather slowly with temperature increase;

(b) *Short wavelengths*, from 2 μm to 8 μm. Although the energy emitted stays very low, it increases sharply when the temperature increases;

(c) *8–13 μm group:* emitted energy is great, even at very low temperatures, and increases 'reasonably,' less suddenly than between 2 and 8 μm, and more sharply than with longer wavelengths.

I drew a boldface line for the 'centers' of the two atmospheric windows, 11 μm and 3.8 μm: we can see the advantage of the spectral band centered on 11 μm, finally chosen for high resolution instruments on Nimbus 4, Nimbus 5, and ERTS B.

Figure 33 is more difficult to read, since I had to use the same reference coordinates for energy values between 0.001 to 300 w/cm^3/ster (spectral radiance). Figure 33 is a detail from the preceding figure, i.e. the emitted energy along narrow spectral bands (50 mμ), the temperature being kept constant at 1000 K (727 °C). The *slope*, very steep with short wavelengths (visible), levels off with near infrared wavelengths. The emitted energy increase coefficients at both extremities of these spectral bands are shown in an inset to the figure.

I think Figure 34 will be of particular interest to remote detection users. I have plotted on the graph the thermal deviations ΔT between real temperatures T and radiance temperatures T_{BB} against emissivity, ε, in the 10.5–12.5 μm band.

When $\varepsilon = 0.98$, ΔT is negligible although greater than 1 °C. With the exception of snow, ice, and unpolluted sheets of water, objects with such high emissivity are rare. Usually ε is very much below 0.95. For $\varepsilon = 0.94$, a fairly commonplace case, $\Delta T = 4$ °C when $T = 0$ °C, and rises to 5 °C when T reaches 40 to 50 °C. When $\varepsilon = 0.80$, which happens for rock formations, soils, ΔT rises from 15 °C ($T = 18$ °C) to 20 °C ($T = 80$ °C).

I plotted these curves to show how illusory and futile is the attempt to find the real temperatures of water, soil, etc. by using radiance temperatures derived from a radiometer (on board a satellite or not).

I assumed a *priori* that the emissivity of a given object remains constant at any temperature whatever. It does not (see end of Part 1, Chapter IV): ε declines with

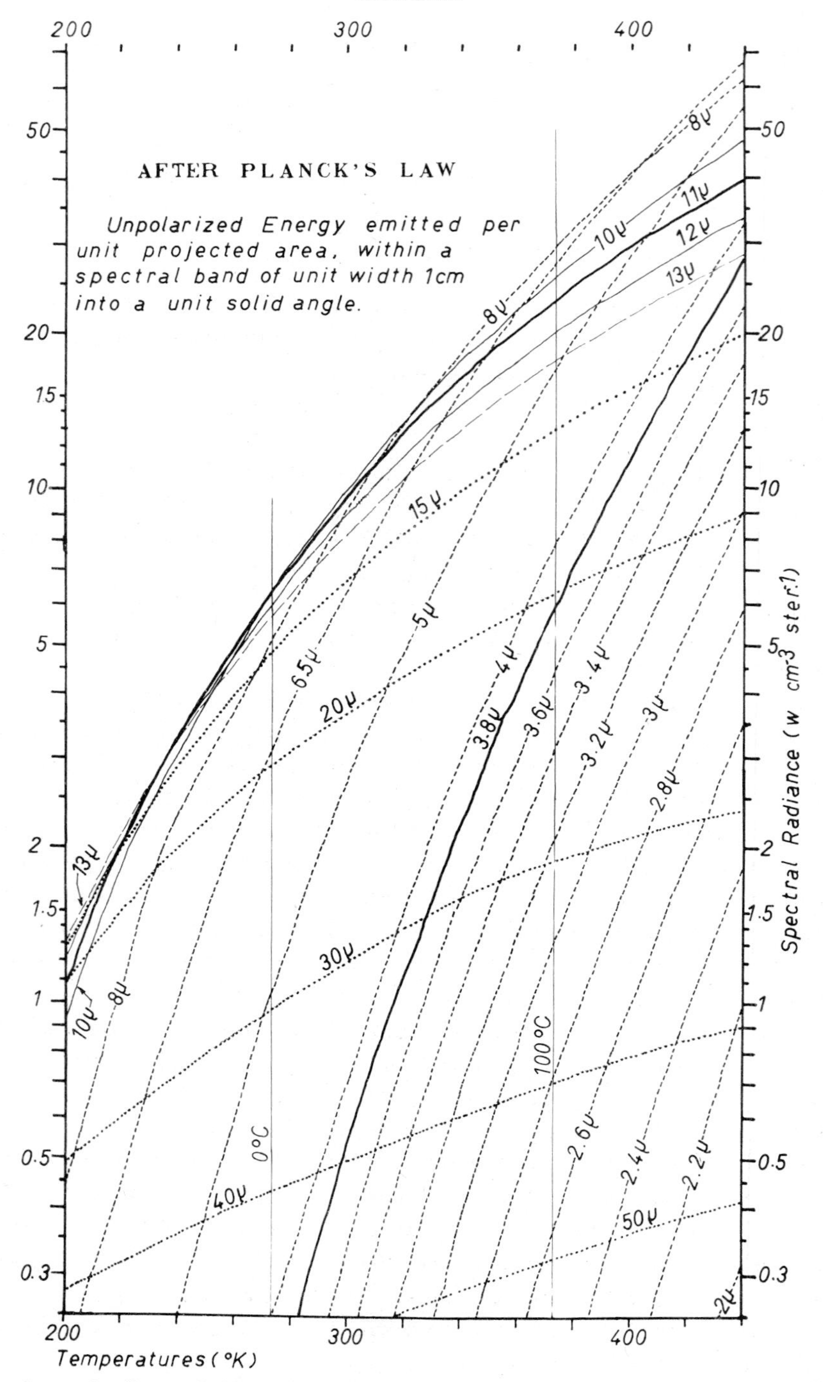

Fig. 32. Spectral radiance of a blackbody at different wavelengths and temperaturss. Unpolarized energy emitted per projected surface unit within a spectral band of unit width 1 cm into a unit solid angle. In watts per cm^3 per steradian.

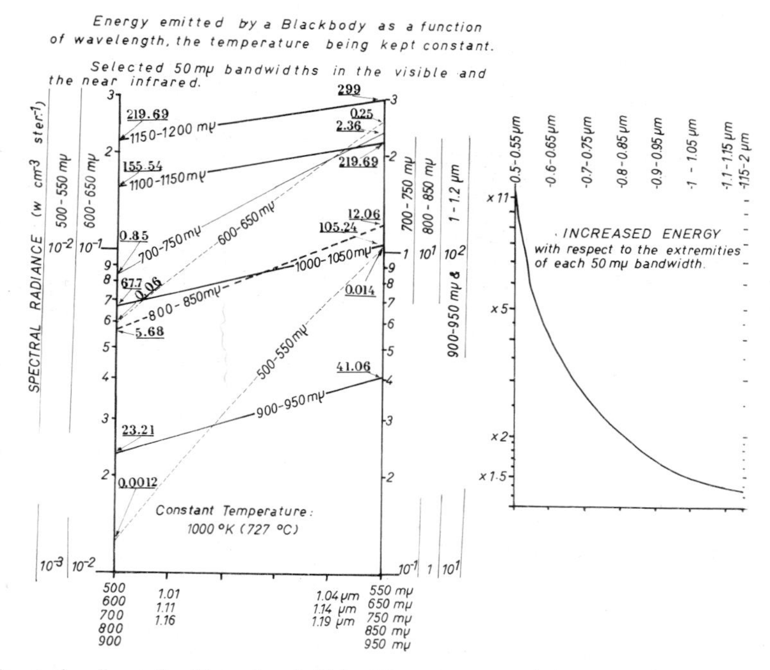

Fig. 33. Spectral radiance for 50 mμ bandwidths. The temperature is kept constant at 1000 K (727 °C). *Inset:* Growth coefficient of emitted energy at the beginning and end of each of the bands selected.

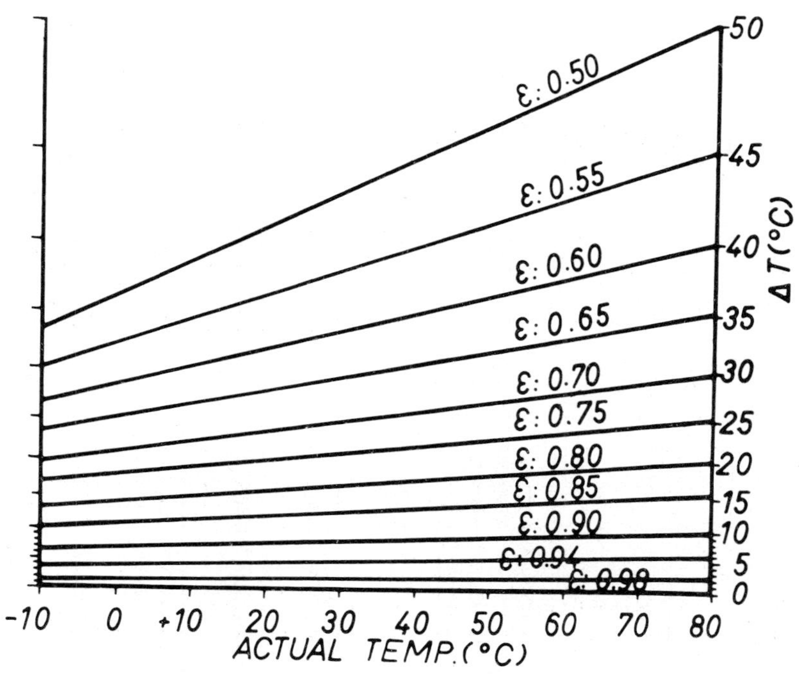

Fig. 34. Temperature deviations with respect to emissivity and real temperatures in the spectral band 10.5–12.5 μm. In all cases the radiance temperature is less than the real temperature.

TABLE XV

Total blackbody radiance

°C	Radiance		°C	Radiance		°C	Radiance		°C	Radiance	
	8–14 μm	10.5–12.5 μm		8–14 μm	10.5–12.5 μm		8–14 μm	10.5–12.5 μm		8–14 μm	10.5–12.5 μm
−30	1.851	0.682	0	3.518	1.215	30	5.762	1.928	60	8.701	2.830
−29	1.993	0.697	1	3.581	1.236	31	5.849	1.955	61	8.805	2.863
−28	2.036	0.712	2	3.644	1.257	32	5.937	1.982	62	8.915	2.896
−27	2.079	0.728	3	3.710	1.278	33	6.026	2.010	63	9.026	2.930
−26	2.123	0.744	4	3.776	1.300	34	6.117	2.038	64	9.138	2.964
−25	2.167	0.760	5	3.843	1.322	35	6.207	2.066	65	9.251	2.998
−24	2.212	0.776	6	3.911	1.344	36	6.299	2.094	66	9.365	3.032
−23	2.257	0.792	7	3.979	1.366	37	6.390	2.123	67	9.479	3.066
−22	2.303	0.808	8	4.047	1.388	38	6.483	2.152	68	9.594	3.100
−21	2.349	0.825	9	4.117	1.410	39	6.575	2.181	69	9.710	3.135
−20	2.396	0.842	10	4.188	1.432	40	6.669	2.210	70	9.827	3.170
−19	2.443	0.859	11	4.260	1.455	41	6.762	2.239	71	9.945	3.205
−18	2.491	0.876	12	4.333	1.478	42	6.856	2.268	72	10.064	3.240
−17	2.540	0.893	13	4.407	1.501	43	6.951	2.298	73	10.184	3.276
−16	2.590	0.910	14	4.482	1.524	44	7.047	2.328	74	10.305	3.312
−15	2.641	0.928	15	4.557	1.548	45	7.144	2.358	75	10.426	3.348
−14	2.693	0.946	16	4.633	1.572	46	7.242	2.388	76	10.548	3.384
−13	2.746	0.964	17	4.709	1.596	47	7.340	2.418	77	10.670	3.420
−12	2.800	0.982	18	4.786	1.620	48	7.439	2.448	78	10.793	3.456
−11	2.855	1.000	19	4.863	1.644	49	7.539	2.478	79	10.916	3.493
−10	2.911	1.018	20	4.941	1.669	50	7.640	2.509	80	11.040	3.530
−9	2.968	1.037	21	5.019	1.694	51	7.742	2.540	81	11.164	3.567
−8	3.026	1.056	22	5.098	1.719	52	7.845	2.571	82	11.289	3.604
−7	3.085	1.075	23	5.178	1.744	53	7.949	2.603	83	11.415	3.641
−6	3.146	1.094	24	5.259	1.769	54	8.054	2.635	84	11.542	3.678
−5	3.206	1.114	25	5.341	1.795	55	8.160	2.667	85	11.670	3.715
−4	3.268	1.134	26	5.424	1.821	56	8.266	2.699	86	11.799	3.753
−3	3.329	1.154	27	5.507	1.847	57	8.373	2.731	87	11.929	3.791
−2	3.392	1.174	28	5.591	1.874	58	8.480	2.764	88	12.060	3.829
−1	3.454	1.194	29	5.676	1.901	59	8.588	2.797	89	12.193	3.867
0	3.518	1.215	30	5.762	1.928	60	8.701	2.830	90	12.327	3.906

increasing T and vice versa. In actual fact, the temperature deviations ΔT are greater than Figure 34 suggests when T is larger than 0 °C, and smaller when T goes below 0 °C.

Tables for temperature-energy relations for spectral bands 8–14 μm and 10.5–12.5 μm (Total blackbody radiation w/cm^2/ster).

Calculations based on Tables II and IV of *Tables of Blackbody Radiation Functions* by M. Pivovonsky, M. R. Nagel and colleagues, Macmillan, New York, N.Y., 1961.

The values serving as reference points are underlined (200°, 260°, 280°, 300°, 320°, 340°, and 360 K).

For the 'directions for use,' remember that $\varepsilon = W_{BB}/W_o$ where ε = emissivity, W_o = = radiance of object, and W_{BB} = radiance of a blackbody at the same actual temperature as the object considered.

The manufacturers (Barnes, Bendix, etc.) supply radiometers operating in the 8–14 μm band; I accordingly used this band for the table, In this case, I advise users conducting field measurements to construct, empirically, a correction table to allow for CO_2 absorption as a function of viewing distance. It must also be remembered that this band has a local minimum (dip) so that ΔT differs according to the silica content of the objects measured.

APPENDIX G

DIFFERENCE OF PARALLAX: HOW TO DERIVE THE HEIGHT OF AN OBJECT FROM AERIAL STEREOSCOPIC PHOTOGRAPHS

1. Preliminary note

When requesting aerial photographs, do not forget to ask for some parameters: altitude of flight; focal length of the cameras; exact size of the negatives (to make sure the prints are not 'blow-ups').

2. Explanation of the symbols used for the equations

H: Flight altitude. Most of the time, it will be advisable to compute this H value from a known distance between landmarks, the distance measured on the photograph, and the focal length.

h: unknown height of a terrestrial feature, bridge, cliff, peak, tree, etc.

B: 'photo-base', that is the distance between the two successive positions of the aircraft. Practically, compute, in millimeters, the distance between the optical center of photo A and the optical center of photo B, reported on the first photograph.

dp: difference of parallax measured with a parallax bar under stereoscopic glasses, assisted by magnifying glasses, dp must be measured in hundredths of millimeters.

3. Figure 35

Point 1: position of the aircraft for the first shot, and *point 2*, position for the second shot.

Let us consider the similar triangles *abc* and *cde*: *set A* where $\overline{ab}=B$, $\overline{de}=dp$; h=height of triangle *cde* and $H-h$=height of triangle *abc*.

This set A does not allow us to derive the value h.

Consequently, drawing parallel lines $\overline{og}$ and $\overline{of}$ drawn from the center of the base $\overline{de}$, we obtain *set B*, 2 similar triangles which are: *ofg* and *cde*. The base of *ofg* is $\overline{fg}=B+dp$ and its height is H. We can write (set B):

$$\frac{h}{H}=\frac{dp}{B+dp}$$

from which we derive the value of h:

$$h=\frac{H\cdot dp}{B+dp}$$

In theory, all the factors, h, dp, H, B should be termed either in km, in m, or in mm.

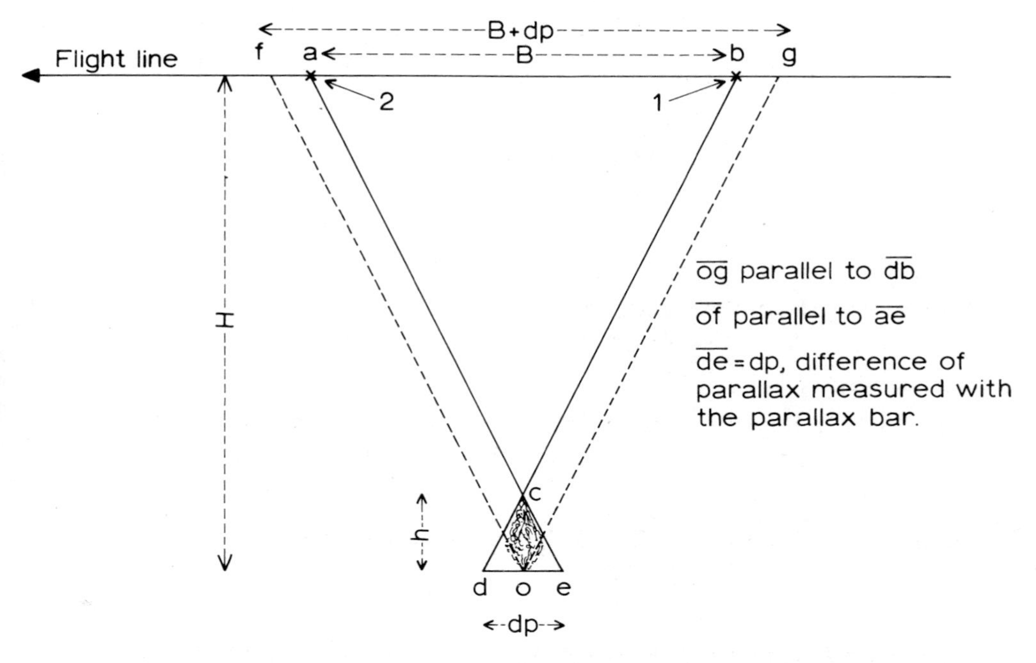

Fig. 35. How to derive the height of an object from the two photographs of a stereogram.

In practice, in order to get rid of too many decimals often leading to miscalculations, H is put in meters and all the other factors in millimeters, the result being in meters.

Ex. $H=3$ km; $B=8$ mm; and $dp=0.04$ mm; in normal units:

$$h=\frac{3000\cdot 0.00004}{0.008+0.00004}=14.92 \text{ m}.$$

in simpler terms:

$$h=\frac{3000\cdot 0.04}{8+0.04}=14.92 \text{ m}.$$

MAP SUPPLEMENTS

Supplement 1. Nighttime radiance temperature distribution, Kalahari region. Nimbus 2, 3.5–4.2 μm radiometer; orbit 1565, 9 September 1966

Supplement 2. Daytime radiance temperatures (10.5–12.5 μm). West Mediterranean. 23 July 1970. Nimbus 4, THIR, orbit 1425.

Supplement 3. Infrared albedo (0.7–1.3 μm), Hoggar region. Nimbus 3, orbit 498, 21 May 1969.

Supplement 4. Daytime radiance temperatures (10.5–12.5 μm). Nile delta-Sinaï area, 19 August 1970. Nimbus 4, THIR, orbit 1787.

(*N.B.* These four maps are offered to the reader for his or her personal interpretation. For maps No. 2 and 4 it is advisable to use graphic symbols to make the maps easier to read.)

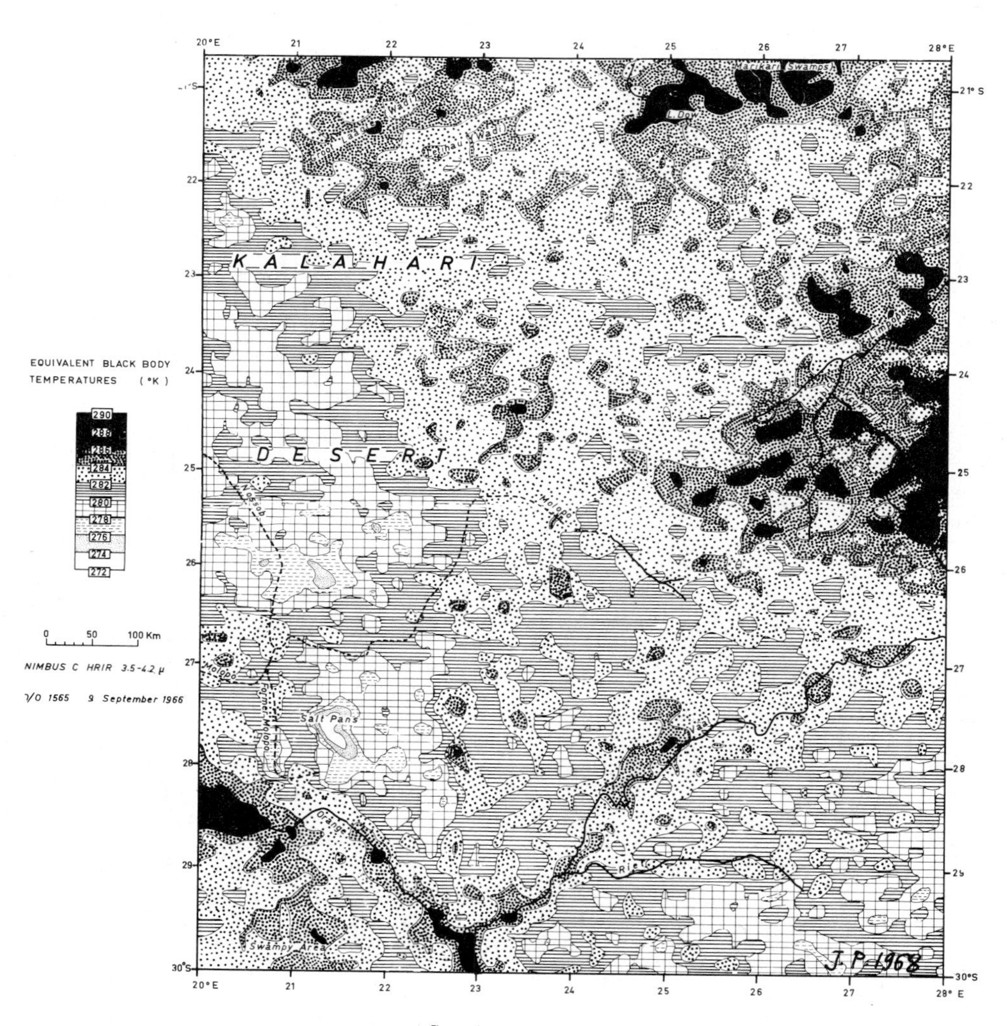

Supplement 1.

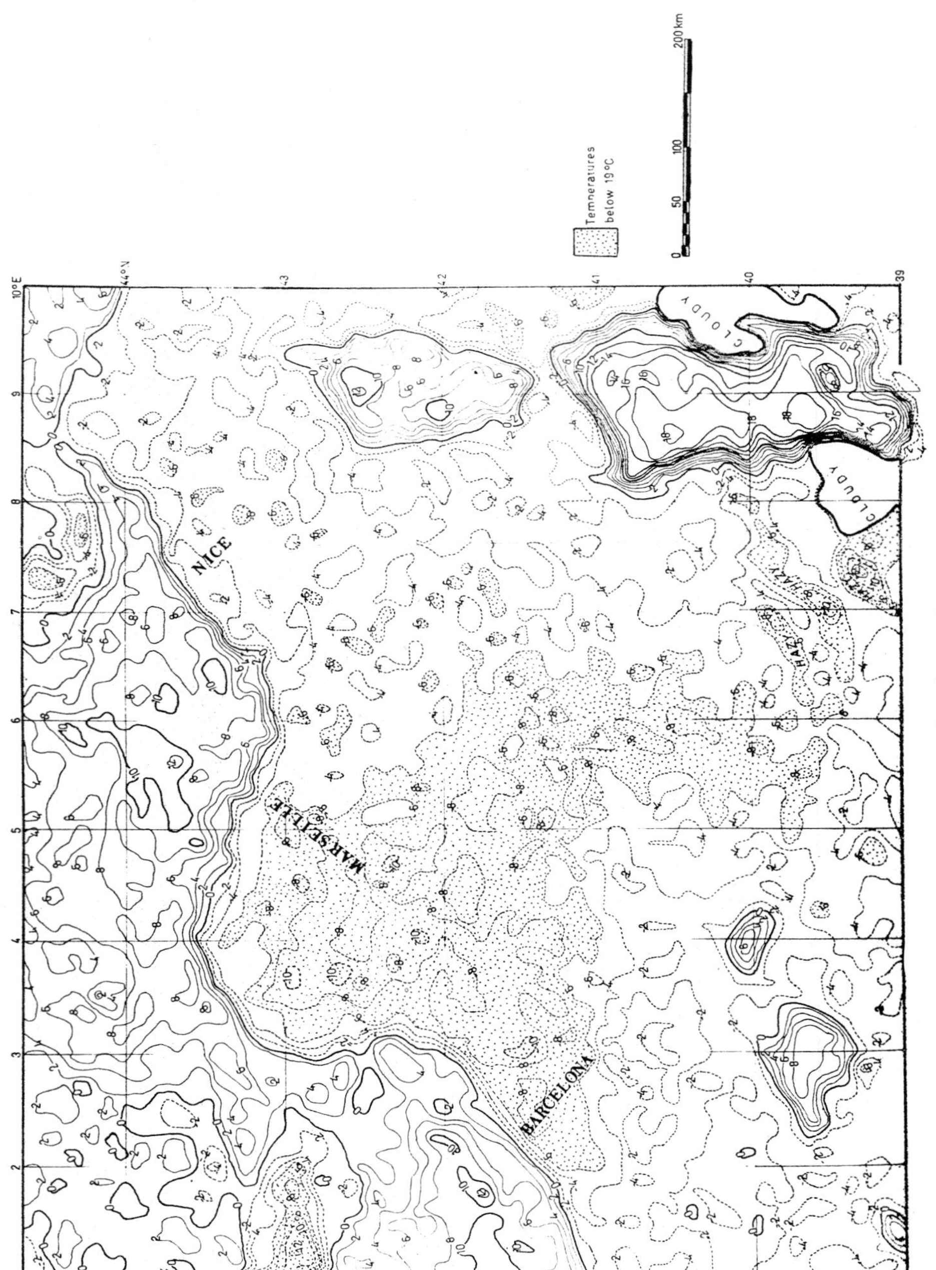

Supplement 2.

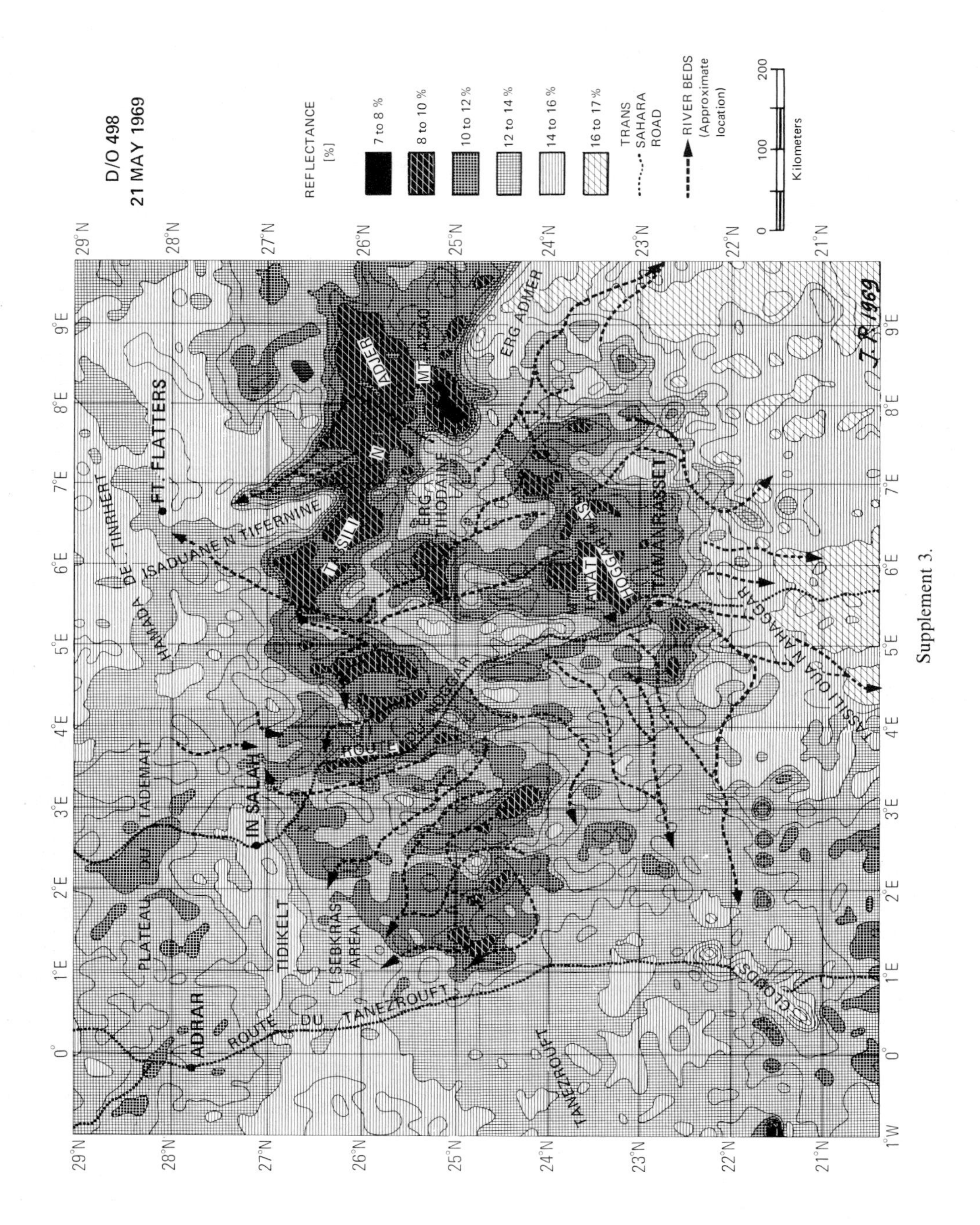

Supplement 3.

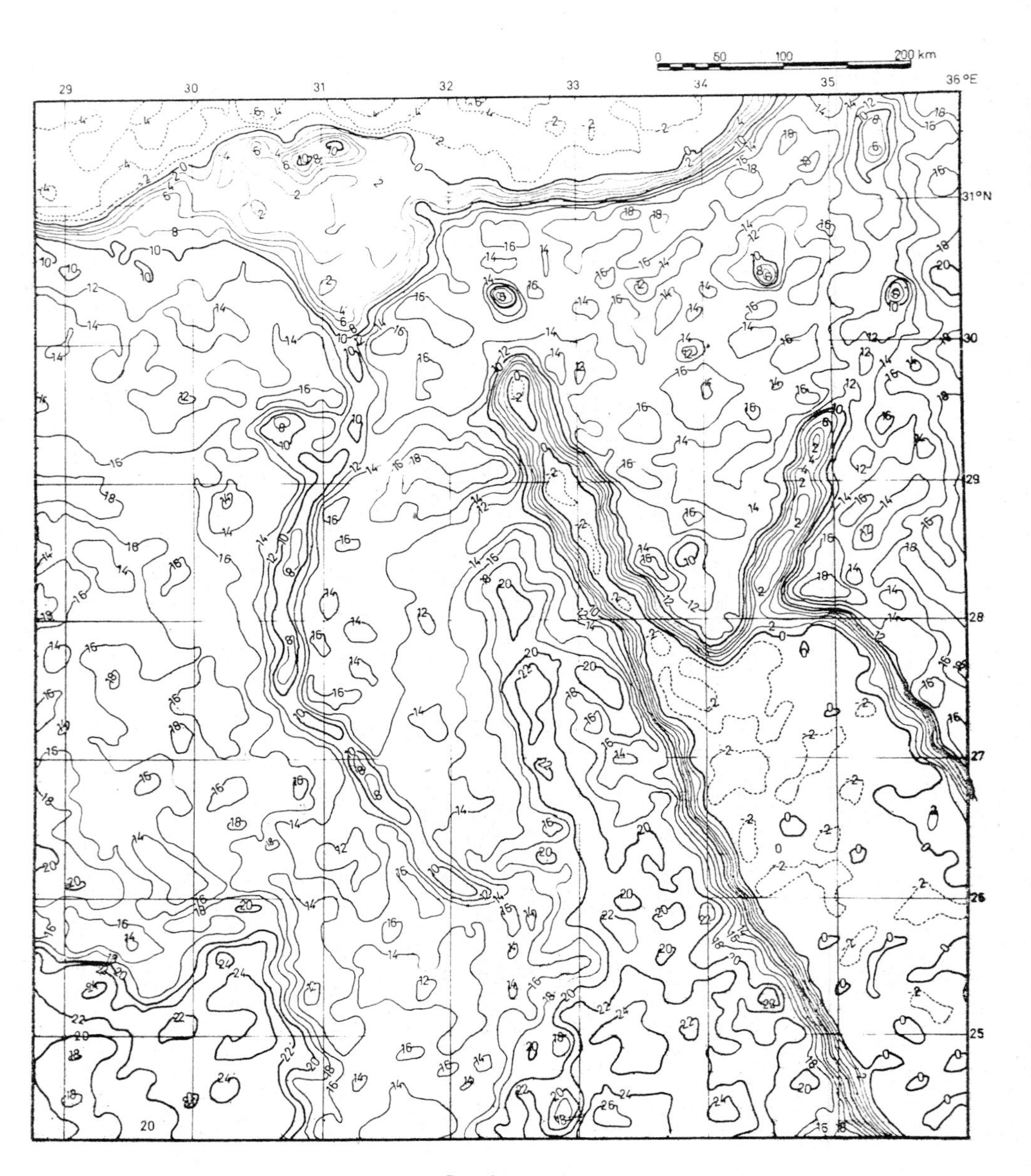

Supplement 4.

BIBLIOGRAPHY

With a very few exceptions I have included here only general articles or books published in 1968. Meteorological publications are mentioned only as an exception.

The following abbreviations will be (unsystematically) used.

AGARD	Advisory Group for Aerospace Research and Development.
AIAA	American Institute for Aeronautics and Astronomy.
Ann.	*Annales, Annals.*
ARA	Allied Research Associates, Inc.
Assn.	Association.
Bull.	*Bulletin.*
CNES	Centre National d'Etudes Spatiales (French 'NASA').
Corp.	Corporation.
ESRO	European Space Research Organization (European 'NASA').
G.B.	Great Britain.
Jour.	*Journal.*
Lab.	Laboratory.
NAS	National Academy of Sciences (Washington, D.C.).
NASA	National Aeronautics and Space Administration.
NRC	National Research Council (Washington, D.C.).
Phot.	Photography.
Photog.	Photogrammetry.
Proc.	*Proceedings.*
Symp.	Symposium.
Un., Univ.	University.
USAF	United States Air Force.
USG	United States Government.
USGS	United States Geological Survey.
Etc.	

1. General Works

Avery, T. E.: 1968, *Interpretation of Aerial Photographs*, 2nd ed., Minneapolis, Minn.

Blackband, W. T. (ed.): 1968, *Advanced Techniques for Aerospace Surveillance*, NATO, for 13th Avionics Symposium, AGARD Panel. Milan, Italy.

Greaves, J. R. and Chang, D. T.: 1970, *Technique Development to Permit Optimum Use of Satellite Radiation Data*, Final Report prepared for Navy Weather Research Facility Detachment, Suitland, Md. ARA, Concord, Mass.

Hubert, L. F.: 1968, *Meteorological Satellites* (McGraw-Hill Yearbook of Science and Technology), McGraw-Hill, Inc., New York.

Johnson, P. L. (ed.): 1969, *Remote Sensing in Ecology*, Univ. of Georgia Press, Athens, Georgia.

NAS-NRC: 1969, *Useful Applications of Earth-Oriented Satellites*, Study Committee Report. Washington, D.C.

NAS-NRC: 1970, *Remote Sensing with Special References to Agriculture and Forestry*, Washington, D.C.

NASA: 1964, 66, 69, 70, 71, 72, *Nimbus User's Guides*, One guide per satellite plus catalogs including photographic mosaics with useful parameters. Except for Nimbus 1, all catalogs are in several volumes. These are sent to official institutions (universities, institutes, etc.) upon application. Write to: Goddard Space Flight Center, Greenbelt, Maryland, 20771, code 601 and code 450.

NASA: *The Application Technology Satellites Meteorological Catalog* (ATS I and ATS III: Vol. I, 1966; Vol. II, 1967; Vol. III, 1968; Vol. IV, 1969; Vol. V, 1970, etc.). For catalogs, write to: Goddard Space Flight Center, ATS Project.
NASA: 1971, *Preparation of Proposals for Investigations Using Data from Earth Resources Technology Satellite (ERTS A) and Skylab (EREP)*. Washington, D.C.
NASA: 1971, *MCS EREP User's Handbook*, Washington, D.C.
NASA: 1971, *Image Formats and Annotations* (relative to ERTS), Washington, D.C.
NASA and General Electric Company: 1971, Earth Resources Technology Satellites. Ground Data Handling System. Preliminary Description. NASA: Washington, D.C. and G.E.: Philadelphia.
Pecker, J. C.: 1969, *L'astronomie expérimentale* (Coll. La Science vivante), P.U.F., Paris.
Smith, J. T. (ed.): 1968, *Manual of Color Aerial Photography*, Amer. Soc. of Photogr.
Tricart, J. *et al.*: 1970, *Introduction à l'utilisation des photographies aériennes*, SEDES, Paris.
Wolfe, W. L.: 1968, *Handbook of Military Infrared Technology*, Univ. of Michigan, Willow Run Lab., Ann Arbor.

2. Specialized Fields

2.1. General

Alouges, A.: see at CNES.
Altshuller, A. P. and McCormick, R. A.: 1970, 'Remote Sensing Platforms for Air Pollution', *J. Remote Sensing* **1**, No. 2.
Barnes, J. C. and Bowley, C. J.: 1968, *Operational Guide for Mapping Snow Cover from Satellite Photography*, ARA, Concord, Mass.
Bennett, G. E. and Hartsook, T.: 1971, *Multispectral Photography (Apollo 9 and Private Flight)*, and *Digitized Map of the Central Rockies Analysis*, Term paper, Geogr. Dept., San Diego State Univ. Copy from NASA Goddard Space Flight Center, Greenbelt, Md.
Buettner, K. J. K.: 1970, *Possible Measurements of Surface Characteristics with Remote Sensors*, XIII COSPAR Meeting, Session A3, Leningrad (provisional publication, Univ. of Washington, Seattle).
Burnett, E. S.: 1970, 'The Earth Resources Observations Satellite Program', *JBIS* **23**, No. 4.
CNES: 1969, 'Les satellites d'étude des ressources terrestres', *La Recherche Spatiale* **8**, No. 5 (author's name not given).
CNES: 1969, 'De Tiros I à Nimbus IV', *La Recherche Spatiale* **9**, No. 5.
CNES: 1971, *La télédétection des ressources terrestres (Remote Sensing of Earth Resources)* ed. by A. Alouges, Paris, CNES.
Colvocoresses, A. P.: March 1970, *Comparison of Basic Modes for Imaging the Earth*, AIAA, Earth Observ. and Inform. Systems, Meeting Annapolis, Md.
Colwell, R. N.: 1968, 'Remote Sensing of Natural Resources', *Sci. Am.* **218**, No. 1.
ESRO: May 1973, 'Directory of European Activities in Remote Sensing of Earth Resources', ESRO, Neuilly sur Seine, France.
Estes, J. E. and Senger, L. W. (eds.): 1972, *An Introductory Reader in Remote Sensing*, Intern. Congr. of Geography, Montreal.
Fischer, W. A.: March 1970, *Projected Uses of Observations of the Earth from Space, The EROS Program of the Dept. of the Interior*, AIAA Earth Res. Observ. and Inform. Systems, Meeting Annapolis, Md.
Gurk, H. H.: 1970, *User Data Processing Requirements*, Typescript from RCA, Astro-Electronics Division.
Hume, C. R.: 1970, 'The *Tiros M* Satellite in an ERTS role', *JBIS* **23**, No. 4.
Keller, D. W.: 1970, 'Earth Resources Satellite Systems', *JBIS* **23**, No. 4.
Laporte, R. and McCarter, W.: 1971, *A Study in Multispectrography and Remote Sensing of the Environment*, Term paper, Geogr. Dept., San Diego State Univ., copy at NASA Goddard Space Flight Center Library, Greenbelt, Md.
McAlister, E. D. and McLeish, W.: 1969, 'Heat Transfer in the Top Millimeter of the Ocean', *J. Geophys. Res.* **74**, No. 13.
Magnolia, L. R.: 1968, 'Selective Bibliography on the Soviet Space Program: 1965–1968', *TRW Space Log* **28**, No. 4.
Newell, H. E.: 1968, 'Current Program and Consideration of the Future Earth Resources Survey', *Space Flight* **10**, No. 8.
Otterman, J. and Bachofer, B. T.: 1970, 'Considerations in Choosing the Orbit for an Earth Resources Survey Satellite', *JBIS* **23**, No. 5.

Palgen, J. J. O.: 1970, *Applications of Optical Processing for Improving ERTS Data*, Technical Report No. 16, Vol. I, ARA, Concord, Mass.

Pecora, W. T.: 1969, 'Surveying the Earth's Resources from Space', *TRW Space Log* **9**. No. 1.

Perry, G. E. and Flagg, R. S.: 1970, 'Telemetry from Russian Spacecraft', *JBIS* **23**, No. 6.

Rabchevsky, G.: 1970, *Remote Sensing of the Earth's Surface*, ARA, Concord, Mass.

Robinove, C. J.: March 1970, *Future Applications of Earth Resources Surveys from Space*, AIAA Earth Res. Observ. and Inform. Systems, Meeting Annapolis, Md.

Sabatini, R. R. and Rabchevsky, G.: 1970, *Use of Ground-Truth Measurements to Monitor ERTS Sensor Calibration*, Technical Report No. 16, Vol. II, ARA, Concord, Mass.

Sheldon, C. S.: 1968, 'The Soviet Space Program, a Growing Enterprise', *TRW Space Log* **28**, No. 4.

Thomas, P. G.: 1968, 'Earth Resources Survey from Space', *Space Aeron.* **50**, No. 1.

Tilden, C. E.: 1968–69, *Techniques for the Utilization of Satellite Observations in Operational Mapping of Sea Surface Temperature*, Meteorology International, Inc., Monterey, Calif. (Quarterly Reports 1, 2, and 3).

TRW Space Log: 1969, 'Summary Log of Space Launches, 1957–1969', **9**, No. 4, pp. 74–133 (author's name not given).

Warren, H. R. *et al.*: 1968, 'Design Considerations for CERES: a Satellite to Survey Canada's Natural Resources', *Can. Aeron. Space. J.* **14**, No. 4.

Wilson, J. E.: 1970, *Sensor Detection Capabilities Study*, USGS, Circular No. 616, Denver, Colo.

Wright, M.: 1971, *Description and Analysis of Apollo 9 Photos and Computer Produced Grid Print Maps of Mesopotamia*, Term Paper, Geogr. Dept., San Diego State Univ., copy from NASA Goddard Space Flight Center, Greenbelt, Md.

2.2. Photographic aspects

Allum, J. A. E.: 1970, 'Assessment of Satellite Photographs for Geological Mapping and Research', *JBIS* **23**, No. 4.

Clark, Ewing, and Lorenzen: 1970, 'Spectra of Backscattered Light from the Sea Obtained from Aircraft as a Measure of Chiorophyll Concentration', *Science* **167**.

Heller, R. C.: 1970, 'Imaging with Photographic Sensors', *Remote Sensing with Special References to Agriculture and Forestry* (see NAS-NRC).

Lowman, P. D.: 1968, *Geological Orbital Photography: Experience from the Gemini Program*, Goddard Space Flight Center, X-644-68-228, Greenbelt, Md.

National Environmental Satellite Center: 1969, *Vidicon Data Limitations*, Technical Memorandum NESCTM 17, U.S. Dept. of Commerce, ESSA, Rockville, Md.

Pease, R. W.: 1970, 'More Information Relating to the High Altitude Use of Color Infrared Film', *Rem. Sens. Envir.* **1**, No. 2.

Pease, R. W. and Bowden, L. W.: 1969, 'Making Color Infrared Film a More Effective High-Altitude Remote Sensor', *Rem. Sens. Envir.* **1**, No. 1.

Rey, P. *et al.*: 1972, Équidensités colorées et Télédétection des Ressources naturelles [Colored Equidensities and Remote Sensing of Natural Resources], CNRS, Service carte de la végétation, Toulouse.

Wobber, F. J.: 1968, *Environmental Studies Using Gemini Photography*, IBM Corp. Technical Report 48-68-002, Bethseda, Md.

2.3. Multispectral

Brennan, B. and Bandeen, W. R.: 1970, 'Anisotropic Reflectance Characteristics of Natural Earth Surfaces', *Appl. Opt.* **9**.

Gausmann, H. W., Allen, W. A., and Cardenas, R.: 1969, 'Reflectance of Cotton Leaves and Their Structure', *Rem. Sens. Envir.* **1**, No. 1.

Hoffer, R. M. and Johannsen, C. J.: 1969, 'Ecological Potentials in Spectral Signature Analysis', in *Remote Sensing in Ecology* (see Johnson, P. L.).

Hovis, W. A.: 1968, *Spectral Studies of Reflected Solar Energy, Snow Fields, Red Tide, Kelp, 0.68 to 2.47 μ C 47 Aircraft*, Goddard Space Flight Center, X-622-68-451, Greenbelt, Md.

Hovis, W. A. and Callahan, W. R.: 1968, *Infrared Reflectance Spectra of Igneous Rocks, Tuffs, and Sandstone from 0.5 to 22 microns*, Goddard Space Flight Center, Greenbelt, Md. (typescript).

Kaltenbach, J. L. (compiled by): 1970, *Apollo 9 Multispectral Photographic Information*, NASA Manned Spacecraft Center, TM-X-1957, Houston, Tex.

Knipling, E. G.: 1969, 'Leaf Reflectance and Image Formation on Color Infrared Film', in *Remote Sensing in Ecology* (see Johnson, P. L.).

Lent, J. D. and Thorley, G. A.: 1969, 'Some Observations on the Use of Multiband Spectral Reconnaissance for the Inventory of Wildland Resources', *Rem. Sens. Envir.* **1**, No. 1.

Lowman, P. D.: 1969, *Apollo 9 Multispectral Photography: Geologic Analysis*, Goddard Space Flight Center, X-644-69-423, Greenbelt, Md.

Planet, W. G.: 1970, 'Some Comments on Reflectance Measurements of Wet Soils', *Rem Sens. Envir.* **1**, No. 2.

Polcyn, F. C., Spansail, N. A., and Malida, W. A.: 1969, 'How Multispectral Scanning Can Help the Ecologist', in *Remote Sensing in Ecology* (see Johnson, P. L.).

Schenkel, F. W.: 1970, 'Design Considerations for Infrared Imaging in the 10–12.6 Micron Band from a Synchronous Altitude Earth Satellite', *JBIS* **23**, No. 6.

Yost, E. and Wendroth, S.: 1969, 'Ecological Applications of Multispectral Color Aerial Photography', in *Remote Sensing in Ecology* (see Johnson, P. L.).

2.4. THERMAL INFRARED AND MICROWAVES

Axelsson, S. and Edwardsson, O.: 1973, 'Side Looking Radar Systems and Their Potential Applications to Earth Resources Surveys', ESRO Publication, CR-71, ESRO, Neuilly sur Seine, France (see also Grant).

Bliamptis, E. E.: 1970, 'Nomogram Relating True and Apparent Radiometric Temperatures of Graybodies in the Presence of an Atmosphere', *Rem. Sens. Envir.* **1**, No. 2.

Dana, R. W.: 1969, *Measurements of 8–14 Micron Emissivity of Igneous Rocks and Mineral Surfaces*, Master's Thesis, Univ. of Washington at Seattle.

Estes, J. E. and Golomb, B.: 1970, 'Oil Spills: Method for Measuring their Extent on the Sea Surface', *Science* **169**, No. 3946 (published by the American Assn. for the Advancement of Science).

Goldberg, I. L.: 1968, *A Very High Resolution Infrared Radiometer Experiment for ATS F and G*, Goddard Space Flight Center, X-622-68-26, Greenbelt, Md.

Grant, K. *et al.*: 1973, 'Side Looking Radar Systems and Their Potential Applications to Earth Resources Surveys', ESRO No. CR-141, ESRO, France.

Greenwood, J. A. *et al.*: 1969, 'Oceanographic Applications of Radar Altimetry from a Spacecraft', *Rem. Sens. Envir.* **1**, No. 1.

Greenwood, J. A. *et al.*: 1969, 'Radar Altimetry from a Spacecraft and its Potential Applications to Geodesy', *Rem. Sens. Envir.* **1**, No. 1.

Haralick, R. M. *et al.*: 1970, 'Using Radar Imagery for Crop Discrimination: a Statistical and Conditional Probability Study', *Rem. Sens. Envir.* **1**, No. 2.

Holter, M. R. *et al.*: 1970, 'Imaging with Non-Photographic Sensors' [radar, passive microwave, infrared, visible, and ultraviolet], in *Remote Sensing with Special References…* (see NAS-NRC).

Hoppe, G. (under the direction of): 1972, 'The Application of Passive Microwave Radiometry to Earth Resources Surveys', ESRO No. CR-75, ESRO, Neuilly sur Seine, France.

Hovis, W. A.: 1968, *Thermal Radiance Spectra 8 to 16 Micron. No. 1, White Sands and the Malpais (lava). C-47 Aircraft*, Goddard Space Flight Center, X-622-68-337, Greenbelt, Md.

Kuers, G.: 1968, *Interpretation of Daytime Measurements by the Nimbus I and II HRIR*, Goddard Space Flight Center, TN-D-4552, Greenbelt, Md.

Myers, V. I. and Heilman, M. D.: 1969, 'Thermal Infrared for Soil Temperature Studies', *Photogram. Eng.* **35**, No. 10.

Nunnaly, N. R.: 1969, 'Integrated Landscape Analysis with Radar Imagery', *Rem. Sens. Envir.* **1**, No. 1.

Waite, W. P.: 1970, 'Snow Field Mapping with K-Band Radar', *Rem. Sens. Envir.* **1**, No. 2.

Walters, R. L.: 1968, *Radar Bibliography for Geoscientists*, Technical Report No. 61–29, University of Kansas, Center for Research and Engineering Science (CRES), Remote Sensing Laboratory.

Weiss, M.: 1968, 'Remote Temperature Sensing', *Oceanology Intern.* **3**, 6.

Williamson, E. J.: 1970, *The Accuracy of the High Resolution Infrared Radiometer on Nimbus II*, Goddard Space Flight Center, TN-D-5551, Greenbelt, Md.

2.5. PRACTICAL RESULTS (EXCEPT METEOROLOGY)*

Auckland, J. C. and Conaway, W. H.: 'Detection of Oil Slick Pollution on Water Surface with Microwave

* Certain publications which involve more than one discipline are classified in Specialized Fields, General, above.

Radiometer System', *Proc. 6th Symp. on Rem. Sens. of Envir.* (published 1970, meeting held 1969).
Auckland, J. C. *et al.*: 1969–70, 'Remote Sensing of the Sea Conditions with Microwave Radiometer Systems', *Proc. 6th Symp. on Rem. Sens. of Envir.*
Burnett, E. S.: 1970, 'Tangible Results Achieved by Earth Observation Satellites to Date', *JBIS* **23**, No. 4.
Dalrymple, R. G.: 1970, *Cartographic Applications of Orbital Photography*, Goddard Space Flight Center, X-644-70-110 (from Gemini 4, 5, 6, 7, 9, 11, and 12 photographs).
Droppleman, J. D.: 1970, 'Apparent Microwave Emissivity of Sea Foam', *J. Geophys. Res.* **75**, No. 3.
Duntley, S. Q. *et al.*: 1970, *Atmospheric Limitations on Remote Sensing of Sea Surface Roughness by Means of Reflected Daylight*, Final Report, Contract NAS-12-2126, NASA, Ames Research Center.
Greaves, J. R. *et al.*: 1968, *Observations of Sea Surface Temperature Patterns and Their Synoptic Changes through Optical Processing of Nimbus II Data*, ARA, Concord, Mass.
Haase, E. *et al.*: 1969, 'Versuch einer Meeresmorphologischen Deutung von Satelliten-Luftbilden', *Deutsche Hydrographische Zeitschrift* **22**, No. 5.
Laing, R. W. and Pardoe, G. K. C.: 1970, 'Remote Observation of Earth Resources', *JBIS* **23**, No. 5.
Merifield, P. M. (ed.): 1969, 'Satellite Imagery of the Earth', *Photogr. Eng.* **35**, No. 7.
Morrison, R. B.: 1970, *Mapping from Space Photography of Quaternary Geomorphic Features and Soil Association in Northern Chihuahua and Adjoining New Mexico and Texas*, Report to NASA, No. 164, Contract No. R-09-020-011, Task No. 160-75-01-46-10 (xerographic typed copies).
Nordberg, W. *et al.*: 1968, *Microwave Observation of Sea State from Aircraft*, Goddard Space Flight Center, X-620-68-44, Greenbelt, Md.
Pouquet, J.: 1968, *An Approach to the Remote Detection of Earth Resources in Arid and Subarid Lands Derived from Nimbus II HRIR Measurements*, Goddard Space Flight Center, TN-D-4647, Greenbelt, Md.
Pouquet, J.: 1969, 'Geomorphology and the Space Age', *Z. Geomorphol., New Series* **13**, No. 4.
Pouquet, J.: 1969, *First Results of Nimbus III: Geological Aspects*, Fall Meeting, American Geophysical Union, San Francisco.
Pouquet, J.: 1969, *Possibilities for Remote Detection of Water in Arid and Subarid Lands Derived from Satellite Measurements in the Atmospheric Window 3.5–4.2 Microns*, Intern. Conference: Arid Lands in a Changing World, Tucson, Arizona.
Pouquet, J.: 1969, 'Les applications géographiques des satellites artificiels', *Bull. Assoc. Géogr. Fr.*, No. 371–372 (notes taken at a conference of the Paris *Institut de Géographie*, written up by the Editorial Staff of the *Bulletin*).
Pouquet, J.: 1970, 'Geopedological Features Derived from Satellite Measurements, in 3.4–4.2 μ and 0.7–1.3 μ Spectral Regions', *Proc. Sixth Intern. Symp....*, Ann Arbor, Mich.
Pouquet, J.: 1972, Résultats de terrain obtenus à l'aide d'un radiomètre portatif opérant dans la bande spectrale 10.5–12.5 μ', *Rev. Geomorphol. Dynam.*
Pouquet, J.: 1970, *Aspects géomorphologiques dans les zones arides et sub arides, d'après les mesures opérées dans les bandes spectrales 0.8–1.3 μm, 3.5–4.2 μm et 10.5–12.5 μm par les satellites Nimbus II, III et IV*, First Intern. Congress on Arid Lands, Mexico City, Mexico.
Pouquet, J.: 1970, Reports to ARA, Concord, Mass.: 1st Report: *The Niger Inland Delta: Situation in 1966 (Nimbus II HRIR) and 1969 (Nimbus III Daytime HRIR)*; 2nd Report: *The Río Paraná-Río Paraguay Area, after Nimbus II and Nimbus III HRIR*; 3rd Report: *First Results Obtained from Nimbus IV THIR*, Channel 11.5 μ (computer-produced grid print map).
Pouquet, J.: 1972, 'Satellites artificiels et cartographie', *Bull. Geogr. Univ. Montreal.*
Pouquet, J.: 1973, First Map of Salton Sea Area Drawn after Passive Microwave Data', *Rev. géogr. phys. geologie dynam.*, Paris (in press).
Pouquet, J. and Raschke, E.: 1968 *A Preliminary Study of the Detection of Geomorphological Features over Northeast Africa by Satellite Radiation Measurement in the Visible and Infrared*, Goddard Space Flight Center, TN-D-4648, Greenbelt, Md.
Resnick, N.: 1970, *Investigation of Small Scale Map Projection for Space Imagery*, USGS, Topogr. Divis., Washington, D.C.
Ribot, J.-M.: 1970, 'L'utilisation en géographie physique des renseignements obtenus par les satellites artificiels', Ph.D. dissertation, 3rd cycle, U.E.R. de Géographie, University of Aix-en-Provence.
Sabatini, R. R. and Sissila, J. E.: 1968 and 1970, *Project NERO (Nimbus Earth Resources Observations)*. 2 vols. ARA, Concord, Mass.
Sissila, J. E.: 1969, 'Observations of an Antarctic Ocean Tabular Iceberg from the Nimbus II Satellite', *Nature* **224**, No. 5226.

Szekielda, K. H.: 1970, *Anticyclonic and Cyclonic Eddies Near the Somali Coast*, Goddard Space Flight Center, X-651-70-265, Greenbelt, Md.
Szekielda, K. H.: 1970, *The Development of the Thermal Structure of Upwelled Water Along the Somali Coast During the Southwest Monsoon.* In manuscript. Goddard Space Flight Center, Code 652, Greenbelt, Md.
Szekielda, K. H. and Laviolette, P. E.: 1970, *Variation in Detailed Sea Surface Temperature Structure Along the Somali Coast as Defined by Nimbus II HRIR Data*, Amer. Geophysical Union Meeting, Washington, D.C.
Tarzwell, C. M.: 1969, Marine Pollution Panel Discussion: Environmental Management in Marine Waters, Proc. Annual Northeastern Regional Antipollution Conference, Univ. of Rhode Island.
USGS: 1968, *Tectonic Map of USSR* (derived from satellite records), Washington, D.C.
USGS: 1970, *Phoenix N1-12-7 Experimental, 1:250000 Scale Space Photomaps. Standard Topographic Line Map, 1:250000 Scale.* (First attempt at cartography using two Apollo 9 photographs superimposed without corrections over the original maps.) USGS, McLean, Va.
Warnecke, G. *et al.*: 1969, *Ocean Current and Sea Surface Temperature Observations from Meteorological Satellites*, Goddard Space Flight Center, TN-D-5142, Greenbelt, Md.
Weaver, D. K. *et al.*: 1969, 'Observations on Interpretation of Vegetation from Infrared Imagery', in *Remote Sensing in Ecology* (see Johnson, P. L.).
Wendland, W. M. and Bryson, R. A.: 1969, 'Surface Temperature Patterns of Hudson Bay from Aerial Infrared Surveys', in *Remote Sensing in Ecology* (see Johnson, P. L.).
Williams, G. F.: 1969, 'Microwave Radiometry of the Ocean and the Possibility of Marine Wind Velocity Determination from Satellite Observation', *J. Geophys. Res.* **74**, No. 18.
Williams, R. S. and Friedman, J. D.: 1970, 'Satellite Observation of Effusive Volcanism', *JBIS* **23**, No. 6.

2.6. Meteorology and Miscellaneous

Allison, L. J. *et al.*: 1969, *Examples of the Usefulness of Satellite Data in General Atmospheric Circulation Research.* Part I, *Monthly Global Circulation Characteristics as Reflected in Tiros VII Measurements*, Goddard Space Flight Center, TN-D-5630, Greenbelt, Md. (See Part II under Godshall.)
Bandeen, W. R.: 1968, *Experimental Approaches to Remote Atmospheric Probing in the Infrared from Satellites*, Goddard Space Flight Center, X-622-68-146.
Bjerknes, J. *et al.*: 1969, 'Satellite Mapping of the Pacific Tropical Cloudiness', *Bull. Amer. Meteorol. Soc.* **50**, No. 5.
Dettwiler, R. H. and Devlin, R. M.: 1970, *Typhoon Cora* (digitized data). Term Paper, Geogr. Dept., San Diego State Univ., copy from NASA, Goddard Space Flight Center, Greenbelt, Md.
Estes, J. E. and Golomb, B.: 1970, 'Monitoring Environmental Pollution', *J. Rem. Sens.* **1**, No. 2.
Godshall, F. E. *et al.*: 1969, *Examples of the Usefulness of Satellite Data in General Circulation Research.* Part II, *An Atlas of Average Cloud Cover Over the Tropical Pacific Ocean*, Goddard Space Flight Center, TN-D-6531 (See Part I under Allison).
Hallgreen, R. E.: 1968, 'The Weather Program', *TRW Space Log* **8**, Nos. 1 and 2.
Hauth, F. F. and Weinman, J. A.: 1969, 'Investigation of Clouds Above Snow Surfaces Utilizing Radiation Measurements Obtained from the *Nimbus II* Satellite', *Rem. Sens. Envir.* **1**, No. 1.
Leese, J. A. *et al.* (eds.): 1969, *Archiving and Climatological Applications of Meteorological Satellite Data*, U.S. Dept. of Commerce, ESSA, Rockville, Md. (Presented in October, 1960 at Geneva, Switzerland, WMO, 5th session, climatology committee.)
McCullough, D. R. *et al.*: 1969, 'Progress in Large Animal Census by Thermal Mapping', in *Remote Sensing in Ecology* (See Johnson, P. L.).
National Environmental Satellite Center: 1970, *Satellite Data Processing Analysis and Interpretation*, U.S. Dept. of Commerce, ESSA, Rockville, Md. (Limited to meteorological aspects.)
Rashchke, E.: 1968, *The Radiation Balance of the Earth's Atmosphere System from Radiation Measurements of the Nimbus II Meteorological System*, Goddard Space Flight Center, TN-D-4859, Greenbelt, Md.
Warnecke, G. and Sunderlin, W. S.: 1968, 'First Color Picture of the Earth Taken from the ATS-3 Satellite', *Bull. Amer. Meteorol. Soc.* **2**.
Woolf, H. M.: 1968, *On the Computation of Solar Elevation Angles and the Determination of Sunrise and Sunset Times.* In manuscript. Goddard Space Flight Center, 1968.

3. Reviews and Periodicals

Only reviews and periodicals that I personally used during my investigations are given.

Bulletin d'information de l'IGN, Paris.
Canadian Aeronautics and Space Journal, Montreal, Que., Canada.
Ciel et Terre, Bulletin of the Belgian Society for Astronomy, Meteorology, and Physics of the Globe, Brussels, Belgium.
JBIS, Journal of the British Interplanetary Society, London, England.
Journal of Remote Sensing, a publication of the International Remote Sensing Institute, Sacramento, California, United States.
La Recherche spatiale CNES (National Center for Space Studies), Paris.
L'Astronomie, Astronomical Society of France, Paris.
Remote Sensing of Environment. An Interdisciplinary Journal, New York, United States.
Spaceflight, publication of the British Interplanetary Society, London, England.
TRW Space Log, publication of TRW Systems Group, Redondo Beach, California, United States.

Because of the regularity of international conferences and the very broad range covered by articles published *in extenso*, I think it useful to classify the following publications among the reviews and periodic bulletins: *Proceedings of the… Symposium on Remote Sensing of Environment*, published after the 1st symposium (1962); 2nd (1963); 3rd (1964); 4th (1966); 5th (1969); 6th (1971)… University of Michigan, Willow Run Laboratories, Ann Arbor, Michigan.